CONTROL OF MACHINES
WITH FRICTION

THE KLUWER INTERNATIONAL SERIES
IN ENGINEERING AND COMPUTER SCIENCE

ROBOTICS: VISION, MANIPULATION AND SENSORS

Consulting Editor: **Takeo Kanade**

ROBOTIC GRASPING AND FINE MANIPULATION, M. Cutkosky
ISBN: 0-89838-200-9

SHADOWS AND SILHOUETTES IN COMPUTER VISION, S. Shafer
ISBN: 0-89838-167-3

PERCEPTUAL ORGANIZATION AND VISUAL RECOGNITION, D. Lowe
ISBN: 0-89838-172-X

ROBOT DYNAMICS ALGORITHMS, F. Featherstone
ISBN: 0-89838-230-0

THREE- DIMENSIONAL MACHINE VISION, T. Kanade (editor)
ISBN: 0-89838-188-6

KINEMATIC MODELING, IDENTIFICATION AND CONTROL OF
ROBOT MANIPULATORS, H.W. Stone
ISBN: 0-89838-237-8

OBJECT RECOGNITION USING VISION AND TOUCH, P. Allen
ISBN: 0-89838-245-9

INTEGRATION, COORDINATION AND CONTROL OF MULTI-SENSOR ROBOT SYSTEMS,
H.F. Durrant-Whyte
ISBN: 0-89838-247-5

MOTION UNDERSTANDING: Robot and Human Vision, W.N. Martin
and J. K. Aggrawal (editors)
ISBN: 0-89838-258-0

BAYESIAN MODELING OF UNCERTAINTY IN LOW-LEVEL VISION,
R. Szeliski
ISBN   0-7923-9039-3

VISION AND NAVIGATION: THE CMU NAVLAB, C. Thorpe (editor)
ISBN   0-7923-9068-7

TASK-DIRECTED SENSOR FUSION AND PLANNING: A Computational
Approach, G. D. Hager
ISBN: 0-7923-9108-X

COMPUTER ANALYSIS OF VISUAL TEXTURES, F. Tomita and S. Tsuji
ISBN: 0-7923-9114-4

DATA FUSION FOR SENSORY INFORMATION PROCESSING
SYSTEMS, J. Clark and A. Yuille
ISBN: 0-7923-9120-9

PARALLEL ARCHITECTURES AND PARALLEL ALGORITHMS FOR INTEGRATED VISION
SYSTEMS, A.N. Choudhary, J. H. Patel
ISBN: 0-7923-9078-4

ROBOT MOTION PLANNING, J. C. LaTombe
ISBN: 0-7923-9129-2

CONTROL OF MACHINES
WITH FRICTION

by

Brian Armstrong-Hélouvry
University of Wisconsin, Milwaukee

KLUWER ACADEMIC PUBLISHERS
Boston/Dordrecht/London

Distributors for North America:
Kluwer Academic Publishers
101 Philip Drive
Assinippi Park
Norwell, Massachusetts 02061 USA

Distributors for all other countries:
Kluwer Academic Publishers Group
Distribution Centre
Post Office Box 322
3300 AH Dordrecht, THE NETHERLANDS

Library of Congress Cataloging-in-Publication Data

Armstrong-Hélouvry, Brian, 1958–
    Control of machines with friction / Brian Armstrong-Hélouvry.
       p. cm. — (The Kluwer international series in engineering and
computer science. Robotics)
    Includes bibliographical references and index.
    ISBN 0-7923-9133-0
    1. Tribology. 2. Machinery—Design. I. Title. II. Series.
TJ1075.A67 1991
621.8′9—dc20   90-20760
                                            CIP

Printed on acid-free paper.

Printed in the United States of America

CONTROL OF MACHINES
WITH FRICTION

Contents

To
Béatrice

Preface

It is my ambition in writing this book to bring tribology to the study of control of machines with friction. Tribology, from the greek for *study of rubbing*, is the discipline that concerns itself with friction, wear and lubrication. Tribology spans a great range of disciplines, from surface physics to lubrication chemistry and engineering, and comprises investigators in diverse specialities. The English language tribology literature now grows at a rate of some 700 articles per year. But for all of this activity, in the three years that I have been concerned with the control of machines with friction, I have but once met a fellow controls engineer who was aware that the field existed, this including many who were concerned with friction. In this vein I must confess that, before undertaking these investigations, I too was unaware that an active discipline of friction existed. The experience stands out as a mark of the specialization of our time.

Within tribology, experimental and theoretical understanding of friction in lubricated machines is well developed. The controls engineer's interest is in dynamics, which is not the central interest of the tribologist. The tribologist is more often concerned with wear, with respect to which there has been enormous progress - witness the many mechanisms which we buy today that are lubricated once only, and that at the factory. Though a secondary interest, frictional dynamics are note forgotten by tribology. In this monograph something over one hundred references to the tribology literature are considered and their implications for control addressed. Interest in frictional dynamics was greatest in the early years of tribological investigation, when the dynamics could be used as a tool to explore basic interface phenomena. Since the 1960's there has been less interest in dynamics, in part because more powerful means have been developed to explore interface physics, and in part because, at least within some quarters of tribology, the dynamics of friction in lubricated machines was considered to be a solved problem. No problem is ever completely solved and, as a half score of recent references attest, interest in dynamics is rising again. This time with predictive models of mechanism friction; models based on three decades of progress in surface and lubricant physics.

ix

For the controls engineer the implications of the modern understanding of friction are substantial. Of the several ways that friction affects machine performance, stick-slip poses perhaps the greatest challenge to precise control. *Stick-slip is dominated by the lubricant characteristics.* All mechanisms are lubricated, even if only by naturally forming oxide films. The lubricanting film can enhance stick-slip, as pure oils or greases will, or it can inhibit stick-slip, as way oil and other special machine oils will. One lubrication engineer, interviewed in the course of this project, went so far as to say that stick-slip could be eliminated in *any* application by appropriate choice of lubricant. The consensus seems to be more cautious; but the point should be well taken that the dynamics of stick-slip, when it occurs, are strongly influenced by the lubricant; and the elimination of stick-slip, where possible, will be greatly facilitated by appropriate lubricant choice. At least a speaking acquaintance with lubrication engineering is important for the controls engineer because the natural priority of the lubrication community is the prevention of wear. It is entirely possible in any particular application that the lubricant was chosen without specific consideration of its anti-stick-slip properties. The experimental research reported here is one example. These experiments were carried out with an industrial robot manipulator using the manufacturer's recommended lubricant. My inquiries led me to the lubrication specialist who had specified the lubricant and who was, quite naturally, proud of achieving a six fold improvement in mechanism life. The stick-slip qualities had, however, received less attention.

In addition to serving as a controls-oriented introduction to the tribology literature, this monograph presents an extensive series of experiments involving friction in a particular machine, culminating in a set of demonstrations of friction compensated motion. There are many specific observations made in the text; but the big observation bears repeating: friction was found to be very highly repeatable. From the standpoint of control implementation this fact is essential, repeatability underlies any form of predictive compensation. From the standpoint of control theory, repeatability, coupled with more exact modeling, opens up the possibility of theoretical results that more accurately reflect the observed phenomena.

This work was made possible by the support, encouragement and wise counsel of my advisors, Professors Thomas O. Binford and Gene F. Franklin; the persistent encouragement and direction of my mentor, Oussama Khatib, and by the availability of facilities at the Robotics Laboratory of the Computer Science Department at Stanford University. Financial support has been provided by the Hewlett Packard Corporation, the Stanford Institute for Manufacturing Automation and the University of Wisconsin-Milwaukee.

The Systron-Donner Corporation contributed markedly to this project by the donation of a rotational accelerometer of exceptional performance. An ounce of sensing may be worth a very great deal of state estimation.

Much has been done since the experimental portion of this work was carried out at Stanford. I am indebted to Professors Joseph D. McPherson and Dilip Kohli, of the University of Wisconsin - Milwaukee; to Susan Rohde, of Northwestern University, and to Carlos Canudas de Wit, of the Laboratoire d'Automatique de Grenoble for their careful reading and many valuable comments. And much is due my parents, Norman and Phyllis Armstrong, who, in more than the evident way, got me started on this undertaking.

I hope and trust that this work will be of use to members of the controls community who are implementing control for machines with friction or who are investigating theory that should rest on a tribological foundation. Friction, as it turns out, is somewhat more complicated, but much less mysterious, than it seems.

Brian Armstrong-Hélouvry
Milwaukee, Wisconsin

Chapter 1

Introduction

Friction is universally present in the motion of bodies in contact. The modern science of tribology seeks to explain the atomic details of friction; but the universality of friction may also be understood from a different perspective. Leonardo da Vinci, among his many investigations, studied the relationship between friction and the music of the heavens. He knew the music to be produced by the bumping and rubbing of the heavenly spheres and was concerned with the possibility of friction between these heavenly bodies:

> "Had however this friction really existed, in the many centuries that these heavens have revolved they would have been consumed by their own immense speed of every day ... we arrive therefore at the conclusion that the friction would have rubbed away the boundaries of each heaven, and in proportion as its movement is swifter towards the centre than toward the poles it would be more consumed in the centre than at the poles, and then there would not be friction any more, and the sound would cease, and the dancers would stop ..."
>
> Leonardo da Vinci (1452-1519),
> *The Notebooks,* F 56 V

The music of the heavens being eternal, Leonardo understood that friction is absent from the state of grace. Thus confined to this mortal world, friction is a consequence of original sin.

1

Friction plays a role in the simpliest actions of living, such as walking, grasping and stacking. In many cases of importance the forces of friction are not small. But for all of this, in the discussion of dynamics for control of mechanical systems friction is but little studied and often completely omitted. When friction is addressed, the models are often those of Leonardo da Vinci or Charles de Coulomb. Experimental evidence pertinent to the situation under scrutiny is rarely sought out or presented. This arises, perhaps, out of the fact that friction may vary from one situation to another, and out of the worth ascribed to problem independent solutions. It seems there is a great impetus toward results that may be applied *sine mutationibus mutandis* to every situation.

The experimental portion of this work is a study of friction in a particular machine: a dc electric robot with spur gears and ball bearings. We shall find along the way a number of effects that extend beyond the confines of this particular mechanism. But more importantly, an example is presented of what may be achieved by mechanism *specific* study. The potential for capturing friction forces in a predictive model is explored; and such fundamental, but long neglected issues as repeatability and structure are addressed directly. The issues addressed in this work - the fraction of friction forces that can be predicted, the model components that are dominant, the characterization of friction at low velocities, a demonstration of experimental procedure for determining basic friction properties - will extend *cum mutationibus mutandis* to a class of mechanisms far broader than that directly studied. Leonardo's conclusions, derived from the study of the motion of bricks on a flat, continue to dominate the design of control when friction is considered. A little experimental work can go a long way.

Robotic force control is the control of contact forces between environment and machine. Force control presents a special challenge that motivates the investigation of friction. This is due, in part, to the unforgiving coupling between actuation and applied force. When the environment and mechanism are very stiff, there is no leeway between the control command and the contact force: a change in control effort is reflected immediately by a change in the contact force. Friction plays a dominant role in limiting the quality of force control. We are accustomed to specifying the upper performance limits of continuously controlled mechanisms: the maximum speed, the greatest motion. The simple theory, even when extended by the classic friction model, fails to predict that there are lower limits. But as a consequence of non-linear low-velocity friction, there are lower limits to motion, a *minimum* speed, a *minimum* distance. These lower limits to motion translate to limits on the fidelity of force control, substantial limits in the case of many practical machines.

A sense of the challenge of force control may be had by considering, for a particular manipulator, the motion corresponding to a small change of force. In a typical configuration the PUMA arm exhibits an end effector stiffness of 20,000 Newton-meters (N-m) per radian. If the desired force resolution is one tenth of a Newton (about one third ounce), at a typical radius of half a meter, it must be possible to govern motions of the mechanism as small as 1/400,000 of a radian. Comparing this to the motion range of the first joint of the PUMA manipulator, which is roughly 5 radians, gives a ratio of desired resolution to total motion of 1 to 2,000,000. If this same requirement were applied to other control systems, it would lead to an elevator for the world's tallest building that could position its car with an error less than the thickness of this page, or a disk head the position of which may be commanded to within a tenth of a wave length of light.

Before proceeding, several definitions are provided - words describing friction are sometimes used imprecisely:

Tribology :
Literally, the study of rubbing; the name given to the modern study of friction and wear of rubbing surfaces.

Static Friction (Sticktion):
The torque (force) necessary to initiate motion from rest. It is often greater than the kinetic friction.

Kinetic Friction (Coulomb friction, Dynamic friction):
A friction component that is independent of the magnitude of the velocity.

Viscous Friction:
A friction component that is proportional to velocity and, in particular, goes to zero at zero velocity.

Break-Away:
The transition from rest (static friction) to motion (kinetic friction).

Break-Away Force (Torque):
The amount of torque (force) required to overcome static friction.

Break-Away Distance:
   The distance traveled during break-away; that is, the distance over which static friction operates, a consequence of the materials used and forces applied.

Dahl Friction or the Dahl Effect:
   A friction phenomenon that arises from the elastic deformation of bonding sites between two surfaces which are locked in static friction. The Dahl effect causes a sliding junction to behave as a linear spring for small displacements.

Stribeck Friction or the Stribeck Effect:
   A friction phenomenon that arises from the use of fluid lubrication and gives rise to decreasing friction with increasing velocity at low velocity.

Negative Viscous Friction:
   Decreasing friction with increasing velocity. Stribeck friction is an example of negative viscous friction.

Deviation:
   Many measurements are reported in this work that exhibit trial to trial variation; deviation is a measure of the trial to trial variation: it is the square root of the sum of the squared variations. Deviation, rather than variance, is used as a measure of variation because the magnitude of the deviation can be compared directly to the magnitude of the signal itself.

   This work addresses a lack of basic experimental data concerning friction in servo-mechanisms. Engineers building feed-forward controllers with friction compensation employ experimental data to adjust model parameters, but no careful investigation of the properties of friction typical of servo-mechanisms has been undertaken. Tribology and the tribological results important for control are surveyed in chapter 2. In chapter 3 the experimental procedures are described. In chapter 4 the repeatability of friction phenomena is examined: any attempt at modeling and compensation is predicated upon repeatability. The motion data show a correlation between measured friction force and position. In chapter 5 this correlation is examined and an experiment that has been highly successful in identifying position-dependent friction is discussed. Chapter 6 addresses the relationships between friction velocity and time; an experiment that provided sensitive measurements of friction at exceptionally low velocities is described. Stick-slip arising from low velocity friction is examined theoretically in chapter 7. Dimensional and perturbation analysis are employed to investigate the conditions under which stick-slip motion will occur. In chapter 8 friction-compensated motion is demonstrated: the friction model is adequate to precompute torques and to move the arm under feed-forward, <u>open-loop</u> control. Cumulative position error during open-loop motions was less than 10%. In the last section of chapter 8 a demonstration of force control is presented. During motion, contact forces are maintained with an error of 0.3% of the level of static friction. Force control fine enough to manipulate wire wrap wire with a PUMA robot. Finally, in chapter 9 the engineering implications of this work are discussed and recommendations presented regarding the design of control for friction-afflicted mechanisms.

Chapter 2

Friction in Machines

In servo controlled machines friction has an impact on the system dynamics in all regimes of operation. Friction, while bemoaned, serves to provide damping at all frequencies, notably those above the bandwidth of control. At the upper limits of performance, friction will affect the design of time optimal control and determine the limits of speed and power. Across the performance spectrum friction contributes to the dynamics. Indeed, in some mechanisms friction may dominate the forces of motion. To accurately design compensation, friction must be modeled.

It is at the low end of the performance spectrum - for tiny motions and corresponding low velocities - that friction posses its greatest challenge. We are accustomed to specifying the upper performance bounds of mechanisms - the greatest displacements or maximum velocity - and friction contributes to the determination of these bounds. But lower performance bounds also exist: a minimum achievable displacement, a minimum sustainable velocity. These lower bounds arise from a periodic process of sticking and sliding, a motion called stick-slip. Stick-slip is generated by the non-linear, low-velocity friction. Stick-slip behavior is distinguished from hunting, which is a periodic oscillation about a fixed goal point. Hunting arises when integral control is used and is often addressed by the use of a dead-band in integral control. Stick-slip may arise during low speed motion with any control design. An applied vibration, called dither, is sometimes used to reduce the impact of friction. Both dither and a deadband in integral control are techniques which run counter to the demands of high-fidelity control.

Friction between dry solid bodies is not a phenomenon involving viscosity and is not described by a viscous friction model. In fact, for a wide range of solid interfaces, friction decreases with increasing velocity. For most machines, however, it is the good fortune of lubrication engineers that full fluid lubrication (no solid body rubbing, the lubricant bears 100% of the

7

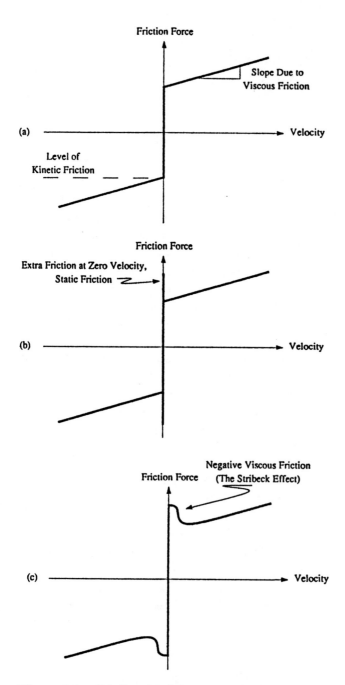

Figure 2.1   Friction Models.   a: Kinetic plus Viscous Friction Model;
b: Static   plus   Kinetic   plus   Viscous   Friction   Model;
c: Negative Viscous plus Kinetic plus Viscous Friction Model.

load) can be achieved. Lubricant is drawn into the solid-to-solid interface by the motion of the parts and maintains a film under pressure through its own viscosity. When there is full fluid lubrication, which will occur in grease as well as oil lubricated mechanisms, a viscous friction model is appropriate. But when motion velocity is low, the motion is insufficient to replenish the lubricant film and solid-to-solid contact results, here a new friction physics begins.

For control of high precision machines, the physics of solid-to-solid contact and the transition from solid-to-solid contact to fluid lubrication pose serious challenges. When parts are in static friction and before motion begins, the displacement allowed by elastic surface deformation can be significant, giving a dynamic for small motions that bears little resemblance to the dynamics of large motions. Friction reversal during velocity reversal is a hard non-linearity. Friction as a function of velocity may be, practically speaking, non-Lipschitz. And the transition from solid-to-solid contact to full fluid lubrication may bring a destabilizing drop in friction. This may occur at velocities that are small but important for tiny motions.

Tribology, the science of friction and wear, has made considerable progress in explaining the atomic details of friction and developing predictive models. But much of the work is applicable to steady state motions, particularly of clean systems. Good experimental evidence collected from practical machines is sparse, but existent, and, in conjunction with theoretical models, provides a basis for more thorough friction modeling. For control, friction as a function of the system state is the quantity of interest. Often friction is considered as a function of velocity. In figure 2.1 common friction models are shown: in figure 2.1(a), a kinetic plus viscous friction model; figure 2.1(b), a static plus kinetic plus viscous friction model; and 2.1(c), a static plus kinetic plus viscous friction model with negative viscous friction at low velocity. Kinetic friction is independent of velocity and is always present. Viscous friction is in proportion to velocity and is present in fluid lubricated junctions, such as machines lubricated by grease or oil. In many situations - though not all - the force required to "break-away", that is to commence motion from rest, is greater than that required to sustain motion, as suggested by the static friction model of figure 2.1(b). Detailed observation of friction suggests that the drop from static friction does not occur instantly and that figure 2.1(c) is often appropriate. The region of negative viscous friction is that in which the friction drops with increasing velocity.

Feedback control is often applied to mechanical arrangements involving metal-on-metal contact and oil or grease lubrication. Issues of manufacture, performance and service life motivate the choice of metals for working members. Grease and oil, which are both fluid lubricants, enormously extend the working life of machine parts and are thus common in servo controlled mechanisms. The reduction of wear achieved by fluid lubrication is so great that for some applications, such as automobile engine journal bearings, more wear may occur in the moments of starting - before the lubricant film is established - than in the hours of operation. For these reasons the use of metal components and fluid lubricants are common, and we may focus on the friction characteristics of these materials when discussing friction and feedback control.

The Classic Models: da Vinci, Amontons, Coulomb and Morin

The classic model of friction - friction force that is proportional to load, opposes the motion, and is independent of contact area - was known to Leonardo da Vinci, but remained hidden in his notebooks for centuries. Rabinowicz [65] argues that the scientific study of friction must have been subsequent to the enunciation of Newton's first law [1687] and the modern conception of force. This is not quite true. Da Vinci's ideas on the nature of force, of which he knew friction to be an example, provide a fascinating insight into problems of pre-Newtonian natural philosophy. Da Vinci viewed force as a sort of ethereal fluid, that attached itself to an object during interaction and sustained motion. He proposed three types of force: a push, an impact and friction. In an intriguing argument, da Vinci concludes that friction is heavier than water, because it sinks to the bottom of rivers and accounts for the bottom water moving more slowly than that on the surface [Da Vinci 1452-1519, The Notebooks].

Da Vinci's friction model was rediscovered by Guillaume Amontons [1699] and developed by Charles Augustin Coulomb [1785], among others. Amontons' claim that friction is independent of contact area (the second of da Vinci's laws) originally attracted skepticism, but was soon verified. Arthur Morin introduced static friction in 1833 and Reynolds the equation of viscous fluid flow in 1866, completing the friction model that is most commonly used in engineering: the Static plus Kinetic plus Viscous friction model [Morin 1883; Reynolds 1886].

Tribology Born

Tribology (Greek for the study of rubbing) was born in England in the 1930's. It was realized that friction consumed vast energy and that wear determined the life of machines. At the time the economic cost of friction was computed to be enormous. This is no less true today. Frank

Phillip Bowden and David Tabor were instrumental in bringing science to the study of friction. Basic questions of wear mechanisms, true contact area, relationships between friction, material properties and lubricating processes were addressed and answered. It is not possible here to give tribology its due. The interested reader is referred to [Bowden and Tabor 56, 73; or Czichos 78], which provide excellent and readable introductions to the field. [Dowson 79] is an engaging work which illuminates the 3,000 year history of man's attempts to understand and modify friction. [Hamrock 86] is a brief handbook survey of the relevant methods of tribology. And [Halling 75] provides a survey that is rigorous but not overly detailed and sufficiently sweeping to address such issues as friction induced instability and solid lubrication. [Ludema 88] is an interesting critique of tribology and cultural barriers to interdisciplinary pursuits.

2.1 The Contemporary Model of Machine Friction

The majority of servo controlled machines are lubricated with oil or grease. Tribologically, greases and oils have more in common than not, grease is essentially a soap matrix that carries oil, releasing the oil under stress to load bearing junctions. These lubricants are widely used because they provide a fluid barrier between rubbing metal parts that exchanges sliding friction for viscous friction and vastly reduces wear. The fluid barrier can be maintained by forcing lubricant under pressure into the load bearing interface, a technique called hydrostatic lubrication. This, however, entails great mechanical complexity and is not applicable to many bearing or transmission designs. The more common technique is that of hydrodynamic lubrication, wherein the lubricant is drawn into the interface by the motion of the parts. Hydrodynamic lubrication is simple to implement, requiring only a bath of oil or grease or perhaps a fluid spray, but suffers the limitation that the fluid film is maintained by motion. Below a minimum velocity the fluid film is lost and solid-to-solid contact occurs.

The Topography of Contact

To understand the tribology of engineering surfaces it is necessary to consider the surface topography. Early models of friction failed because the surface topography was misunderstood. The interactions at contacting surfaces will be examined by considering progressively smaller contacts. In figure 2.2 a conformal contact is shown schematically; part A rests on part B. Kinematically, such contacts are identified as area contacts: the apparent area of the contact is determined by the size of the parts.

Conformal Contact

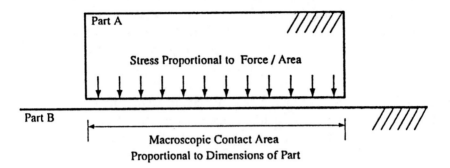

Figure 2.2   Conformal Contact, such as Machine Guide Ways or Journal Bearings.

Nonconformal Contact

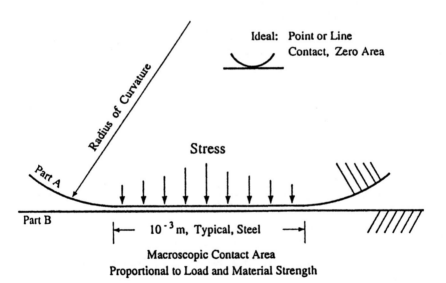

Figure 2.3   Nonconformal Contact, such as a Gear Tooth Mating or Roller Bearings.

Parts that do not enjoy a matching radii of curvature meet at nonconformal contact, as shown in figure 2.3. These contacts are called point or line contacts when considered kinematically; but this is an idealization.

True Contact Between Engineering Surfaces

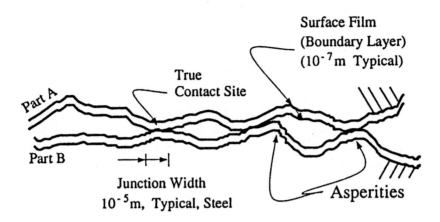

Figure 2.4 Part to Part Contact Occurs at Asperities, the Small Surface Features.

In fact the parts deform to create an apparent area of contact, an area that increases with increasing load. The one milli-meter contact width suggested in figure 2.3 is typical of small machine parts, such as the transmission gears of an industrial robot.

Tribology as a field is sophisticated in the use of similitude. One transformation that is widely used maps a nonconformal contact of two radii to one of a flat surface and a single curved part, as suggested in figure 2.3 [Dowson and Higginson 66; Hamrock 86]. This transformation greatly simplifies the study of nonconformal contacts. Nonconformal contacts are often called hertzian contacts, after the original analysis [Hertz 1881], and arise frequently in machinery. The stresses found in conformal contacts between steel parts are rarely higher than 7 mega-Pascals (MPa), whereas in nonconformal contact the peak stress can be 100 times greater [Hamrock 86]. A stress of 700 mega-Pascals corresponds to 100,000 psi, which is greater than the yield strength of many types of steel. This is possible in hertzian contact because the stress is compressive.

In a BBC radio program, F. P. Bowden [50] observed that ".. putting two solids together is rather like turning Switzerland upside down and standing it on Austria - the area of intimate contact will be small." Crystalline surfaces, even apparently smooth surfaces, are microscopically rough. The protuberant features are called asperities and, as shown schematically in figure 2.4, the true contact occurs at points where asperities come together. In this way the true area is much smaller than the apparent area of the contact [Bowden and Tabor 39]. Over a broad range of engineering materials, the asperities will have slopes ranging from 0 to 25 degrees and concentrated in the band from 5 to 10 degrees [Dowson 79].

When asperities come into contact the local loading will be determined by the strength of the materials, which will deform as necessary to take up the total load. As a first approximation, we may consider the local stress at an asperity junction to be in proportion to the yield strength of the material. The contact area, on the other hand, is in direct proportion to the total load. As a rule of thumb, the true contact area, A, is given by $A = W/3Y$, where W is the load and Y is the yield strength of the material. Contact stress at the asperity is taken, by this rule of thumb, to be three times the yield strength. As with the nonconformal contact, stress greater than yield strength is possible because the asperities are under compression.

Friction is proportional to the shear strength of the asperity junctions. As the load grows, the junction area grows; but, to first order, the shear strength (measured per unit area) remains constant. In this way friction is proportional to load. If truly clean metal surfaces are brought into contact, the shear strength of the junction (friction) can be the shear strength of the bulk material, and the friction coefficient can be much greater than one [Bowden and Tabor 73; Hamrock 86]. Fortunately for the operation of machines, this is all but impossible to achieve. Even in the absence of lubricants, oxide films will form on the surface of steel and other engineering materials, producing a boundary layer. In the presence of lubricants, additives to the bulk oil react with the surface to form the boundary layer. These additives are formulated to control the friction and wear of the surface. The boundary layer is a solid, but because it has the lower shear strength, most shearing occurs in this film. If the boundary layer has a low shear strength, friction will be low; if it has good adhesion to the surface and can be replenished from the oil, wear will be reduced. Boundary layer thickness varies from a few atomic thicknesses to a fraction of a micron. As suggested in figure 2.4, a tenth of a micron is a typical thickness of the boundary layer formed by the lubricity additives of industrial oil [Wills 80; Booser 84]. Note that this is perhaps two orders of magnitude less than the typical dimension of an asperity in steel junctions. The boundary layer is exactly that, like top soil in Austria, and does not markedly influence the area or local stresses of contact.

Friction as a Function of Velocity: Four Dynamic Regimes

There are four regimes of lubrication in a system with grease or oil: static friction, boundary lubrication, partial fluid lubrication and full fluid lubrication. These four regimes each contribute to the dynamic that a controller confronts as the machine accelerates away from zero velocity. Figure 2.5 is known as the Stribeck curve and shows the three moving regimes [Stribeck 02; Biel 20; Czichos 78]. The interesting characteristics of static friction are not dependent on velocity.

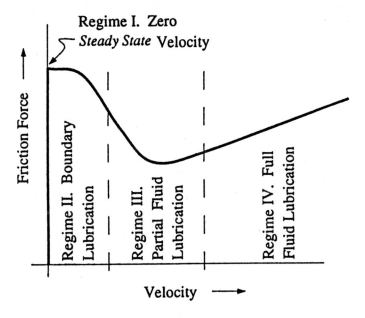

Figure 2.5 The Generalized Stribeck Curve, showing Friction as a Function of Velocity for Low Velocities.

The First Regime: Static Friction

In figure 2.4 contact is shown to occur at asperity junctions. From the standpoint of control, these junctions have two important behaviors: they deform elastically, giving rise to motion that appears to be a solid connection with a stiff spring; and both the boundary film and the asperities deform plastically under the load, giving rise to increasing static friction as the junction spends more time at zero velocity.

It is often assumed that there is no motion while in static friction, which is to say no motion without sliding. Dahl [68, 76, 77], studying experimental observations of friction in small rotations of ball bearings, concluded that a

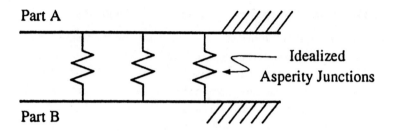

Figure 2.6 Idealized Contact Between Engineering Surfaces in Static Friction. Asperity Contacts Behave Like Springs.

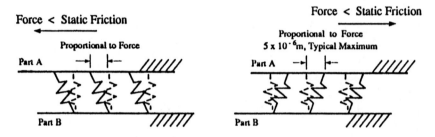

Figure 2.7 Asperity Deformation under Applied Force, the Dahl Effect.

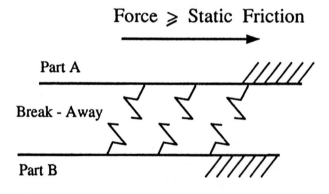

Figure 2.8 At Break-Away True Sliding Begins.

junction in static friction behaves like a spring. That is, the force is a linear function of the displacement, up to a critical displacement (force) at which break-away occurs. The elasticity of asperities is suggested schematically in figure 2.6. When forces are applied, the asperities will deform, as suggested by figure 2.7, but recover when the force is removed, as does a spring. In this regime, friction is governed by:

$$F(x) = -kx \; ; \tag{2.1}$$

where k is the effective stiffness of the asperity contacts and x is displacement. When the applied force exceeds the level of static friction, the junctions break (in the boundary layer, if present) and true sliding begins [Gassenfeit and Soom 88]. This is suggested in figure 2.8. This phenomenon, termed the Dahl effect [Armstrong 88], is considered by Walrath [84] in an adaptive controller design. Villanueva-Leal and Hinduja [84] and Cheng and Kikuchi [85] have examined the Dahl effect using finite element techniques and contact models. Their results agree with experiment and point the way toward predictive models. [Burdekin, *et al.* 78] examine the shear stiffness of contacts between cast iron, and find that it is strongly influenced by the normal force.

The distance over which the Dahl effect operates may be minute, break-away is observed to occur with deflections on the order of 5 microns in steel junctions [Rabinowicz 51; Dahl 68; Burdekin, *et al.* 78; Cheng and Kikuchi 85; Villanueva-Leal and Hinduja 84]. But elsewhere in a mechanism a much greater displacement may be observed, displacements significant on the scale of feedback control. This will arise, for example, in robots, where the arm itself acts as lengthy lever.

The Second Regime: Boundary Lubrication

In the second regime fluid lubrication is not important, the velocity is not adequate to build a fluid film between the surfaces. As described, the boundary layer serves to provide lubrication; it is solid, but of lower shear strength to reduce friction. In figure 2.9 sliding in boundary lubrication is shown. Because there is solid-to-solid contact, there is shearing in the boundary lubricant. Because boundary lubrication is a process of shear in a solid, it is often assumed that friction in boundary lubrication is higher than for fluid lubrication, regimes three and four. This, however, is not always the case; it is not necessary that the shear strength of a solid be greater than the viscosity of a fluid. Consider the viscosity of glass: many solids will yield to a lower shear force than the forces of viscous flow in this fluid. Certain boundary lubricants do reduce static friction to a level below kinetic friction and entirely eliminate stick-slip. Some of the chemistry of these and other boundary lubricants are described is section 2.2.

Boundary Lubrication, Regime II of the Stribeck Curve (Low Velocity True Sliding)

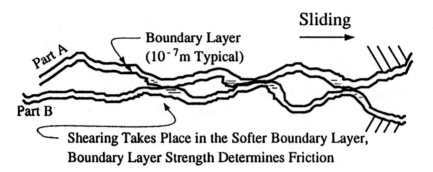

Figure 2.9   Sliding with Solid-to-Solid Contact Causes Shearing in the Boundary Layer.

The Third Regime: Partial Fluid Lubrication

Shown in figure 2.10 is the process by which lubricant is drawn into the contact zone, in this case of a nonconformal contact. Lubricant is brought into the load bearing region through motion, either by sliding or rolling. Some is expelled by pressure arising from the load, but viscosity prevents all of the lubricant from escaping and thus a film is formed. The entrainment process is dominated by the interaction of lubricant viscosity, motion speed and contact geometry. The greater viscosity or motion velocity, the thicker the fluid film will be. When the film is not thicker than the height of the asperities, some solid-to-solid contact will result and there will be partial fluid lubrication. When the film is sufficiently thick, separation is complete and the load is fully supported by fluid.

Partial fluid lubrication is shown schematically in figure 2.11. The dynamics of partial fluid lubrication can perhaps be understood by analogy with a water skier. At zero velocity the skier is supported buoyantly in the water. Above some critical velocity the skier will be supported dynamically by his motion. Between floating and skiing there is a range of velocities wherein the skier is partially hydrodynamically supported. These velocities are analogous to the regime of partial fluid lubrication. The analogy is imperfect in that the buoyant support is not like solid-to-solid contact; and the dynamic support of the skier is due to fluid inertia as opposed to viscosity, the dominant force in lubrication. In one aspect, however, the

Lubricant Entrainment by Motion

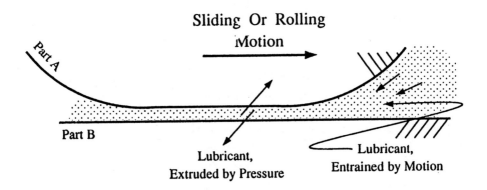

Figure 2.10   Motion Brings Fluid Lubricant into the Contact Zone.

Partial Fluid Lubrication, Regime III of the Stribeck Curve (Moderate Velocity True Sliding)

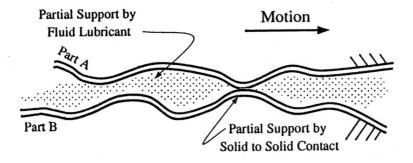

Figure 2.11   Partial Fluid Lubrication, the Fraction of the Load Supported by Fluid Lubricant is Velocity Dependent.

analogy is valid: for both the water skier and the machine, the regime of partial dynamic support is manifestly unstable.

Partial fluid lubrication is the least well understood of the four regimes. In the case of nonconformal contact, even full fluid lubrication (Elasto-Hydrodynamic Lubrication, or EHL) must be investigated numerically. For these contacts, steady state flows over smooth surfaces are well understood

[Dowson and Higginson 66; Booser 84; Pan and Hamrock 89]; but these are not the true conditions of partial fluid lubrication. Work is proceeding toward an understanding of the interaction of surface roughness and EHL in steady state motion [Zhu and Cheng 88; Sadeghi and Sui 89]. From these papers it appears that the details of surface roughness, asperity size and orientation, have significant impact on the lubricant film characteristics, complicating a general analysis.

Of principal interest to the controls engineer is the dynamics of partial fluid lubrication with changing velocity. Theoretical study of this problem is beginning [Sroda 88; Rayiko and Dmytrychenko 88]. These numerical investigations show a time delay between a change in the velocity or load conditions and the change in friction to its new steady state level. This delay, called frictional lag, has been observed experimentally in a wide range of circumstances [Rabinowicz 58; Bell and Burdekin 69; Rice and Ruina 83; Walrath 84; Hess and Soom 90]. The delay may be on the order of milli-seconds to tens of milli-seconds, and its impact on stick-slip motion, considered below, may be substantial. As a delay in the appearance of a destabilizing drop in friction, frictional lag may in fact be a boon to stable control.

The Fourth Regime: Full Fluid Lubrication, Hydrodynamic or Elasto-Hydrodynamic

Hydrodynamic and elastohydrodynamic lubrication (EHL) are two forms of full fluid lubrication. Hydrodynamic lubrication arises in conformal contacts, and EHL in nonconformal contacts. As figure 2.12 shows, solid-to-solid contact is eliminated. In this regime wear is reduced by orders of magnitude and friction is well behaved. The object of lubrication engineering is often to maintain full fluid lubrication effectively and at low cost. Reynolds [1886] and Sommerfeld [04] laid the ground work for the investigation of hydrodynamic lubrication, which has been worked out in great detail (see, for example, [Hersey 14, 66; Halling 75]).

EHL is common in servo controlled machines. As mentioned, it is studied numerically: there is no analytic solution simultaneously satisfying the surface deformation and fluid flow equations. Generally speaking, EHL will give higher friction and wear than hydrodynamic lubrication, as suggested by figure 2.13.

General predictive models of the steady state lubricant film thickness are available [e.g., Halling 75; Hamrock 86]. The film thickness, which determines friction as well as protection from wear, is a function of surface rigidity and geometry, lubricant viscosity and velocity. The principal value

Full Fluid Lubrication, Regime IV of the Stribeck Curve (High Velocity True Sliding)

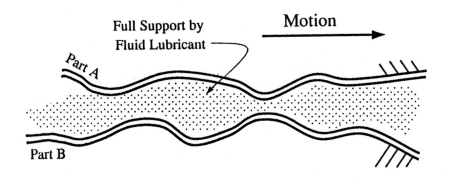

Figure 2.12   Full Fluid Lubrication, No Solid-to-Solid Contact.

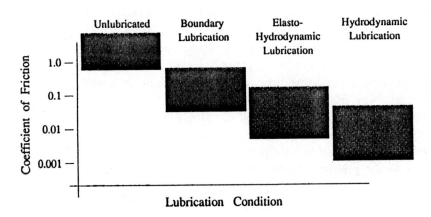

Figure 2.13   The Range of Friction Levels.

for control in these results will be in predicting the transition velocity to full fluid lubrication. Work is beginning in the exploration of the transient dynamics of elastohydrodynamic lubrication [Xiaolan and Haiqing 87].

2.2 Boundary Lubricants: a Domain of Many Choices

Boundary lubrication is important to the controls engineer because of the role it plays in stick-slip. The key to effective boundary lubrication is the discovery of a molecule that binds with reasonable strength to the metal surface, but is not corrosive; that has sufficient strength to withstand the forces of sliding and yet has a low shear strength to give low friction. Such molecules are added to the bulk lubricant, often comprising only a percent or two of the total. Lubrication additives may be broken into three broad classes:

- Lubricity Agents;
- Extreme Pressure Agents;
- Anti-Wear Agents.

Long chain hydrocarbons with a polar group at one end are commonly used as lubricity agents. The polar group bonds to the metal and the long chain sticks away from the surface, creating, in effect, a mat of bristles [Merchant 46; Bowden and Tabor 73; Fuller 84]; the longer the chain (bristle) the lower the friction. These additives are sometimes called oiliness agents, anti-friction agents or friction modifiers. Friction modification refers to reducing the static friction and friction in boundary lubrication. The polar hydrocarbons attach themselves to the metal surface by charge exchange in a process called "physi-adsorption". Their application is limited to situations of moderate temperature. At approximately 100 °C the polar hydrocarbons desorb (detach) and boundary lubrication is lost [Bowden and Tabor 73; Fuller 84]. For this reason the use of long chain hydrocarbons is restricted to applications that generate relatively little frictional heating, which is generally a restriction to conformal contacts.

Use of these polar hydrocarbons as friction modifiers is wide spread in the form of "way oils", oils specially formulated to eliminate stick-slip in machine slideways [Merchant 46; Wolf 65; Mobil 78]. Machine slideways are conformal, and thus less affected by frictional heating. A premium is placed on eliminating stick-slip in precision machine tools and great attention has been give to the problem [Merchant 46; Wolf 65; Bell and Burdekin 66, 69; Kato *et al.* 72]. The level of static friction can, in fact, be reduced below the level of kinetic friction so that there is no destabilizing negative viscous friction and stick-slip is eliminated [Merchant 46; Wolf 65; Mobil 78; Wills 80]. There are standard procedures for measuring the lubricity of ways oils, one is the Cincinnati Milacron Stick-Slip Test [Cincinnati Milacron 86]. This test procedure measures the friction at break-away and at a velocity of 0.5 inches per minute. The Cincinnati Milacron test procedure is quite similar to that described in [Wolf 65]. A static friction level that is less than 0.85 of the kinetic friction level ($F_s/F_k \leq 0.85$) indicates that stick-slip will be eliminated [Wolf 65]. F_s/F_k values as low as 0.55 are observed [Millman 90]. The possibility of wider use of these lubricants in servo machinery is an intriguing one.

Extreme pressure (EP) agents chemically react with the metal to form a film that will protect the surface from wear. The principle issue in their formulation has been the reduction of wear and seizure [Papay 74, 76, 88], but most EP additives also provide a degree of friction modification and some will pass standard stick-slip tests [Facchiano 84; Lubrizol 88; Cincinnati Milacron 86]. EP agents are available in a vast variety, and are universally present in gear and other machine lubricants and thus in many servo controlled machines. EP agents bond with the surface by chemically reacting with the metal, or "chemi-adsorption". For this reason they tend to be metal specific. EP additives function at higher temperatures than do lubricity additives and so are serviceable under more severe loading, such as in nonconformal contacts. The chemi-adsorption also offers a generally stronger bond to the surface and thus greater protection against wear. The principle limitations of EP additives are a weaker friction modification than is achieved by the lubricity agents, and chemical reaction with the surface, which is by its nature corrosive. With EP agents one in effect acquires greatly reduced mechanical wear at the price of slow corrosion [Wills 80; Fuller 84; Papay 88].

Anti-wear agents extend the service life of machine parts through a remarkable chemistry that can repair some forms of wear induced surface damage [Estler 80; Booser 84]. The issue of principle concern for the controls engineer is that the anti-wear agent can interfere with the friction modification of lubricity or EP additives. Lubricant additives also perform a host of other functions, including viscosity modification, foam control, corrosion protection, and oxidation stabilization [Papay 88]. These functions are key to machine and lubricant life, but do not bear directly on mechanism dynamics or control.

Lubricant additives must stay in suspension or solution in the bulk lubricant. In this way they are available to replenish sites on the surface where the lubricant film is damaged by rubbing [Bowden and Tabor 73; Hamrock 86]. Replenishment of the boundary layer from the bulk lubricant may be required after each pass [Vinogradov *et al.* 67; Cameron 84]. Boundary lubricants are standard additives in machine grease or oil; there is a great range of formulations, and they typically constitute less than 2% of the total. Systems with high loading and low relative velocity, such as gear teeth, may operate entirely in boundary lubrication [Mobil 71; Wellauer and Holloway 76; Wilson 79]. Much of the attention in boundary lubricant formulation has been focused on reduction of wear. In the design of lubricants other than way lubricants, friction modification has often played a secondary role.

Dry lubricants, such as teflon, operate by a variety of mechanisms. Their principal liability is the loss of the protection against wear provided by full fluid lubrication. A good survey of dry lubrication may be found in either [Halling 75] or [Fuller 84]. From a control perspective, dry lubricants may offer the substantial advantage of eliminating the negative viscous friction associated with partial fluid lubrication. stick-slip.

2.3 Relaxation Oscillations

For fine machine motion, it is the stick-slip phenomenon that poses the greatest challenge. Stick-slip was apparent in early studies of low speed motion. The first attempts at explanation were carried out within the static plus kinetic friction model of figure 2.1(a) [Thomas 30]. Using a sensitive displacement measuring apparatus, photomicrographs of the rubbing surfaces and hydraulically produced steady motion, Bowden and Leben [39] demonstrated that sticking occurs and coined the term stick-slip [Rabinowicz 56a]. They observed welding in the photomicrographs and, using the thermocouple effect between dissimilar metals, they found wide temperature fluctuations that are correlated with the stick-slip cycle. Bowden and Leben [39] posited local melting of one rubbing metal as a mechanism for decreased friction during sliding. They found that a similar stick-slip occurs in many lubricated systems, even if there is no welding; and that no stick-slip occurs when long chain fatty acids are used as a lubricant. At the time boundary lubricants were not well understood. The fatty acids used by Bowden and Leben, [39], are now commonly used as lubricity agents.

By 1940 experiments had not been conducted which could observe the details of friction during a stick-slip cycle. But it became evident from macroscopic observations, in particular the range of speeds and structural conditions over which stick-slip will occur, that the static plus kinetic friction model was inadequate to explain the observed phenomena. Dudley and Swift [49] employed phase plane analysis to study the possible oscillations in slider mechanisms, that is mass spring systems with no feedback control. A negative viscous friction was posited and efforts were directed at elucidating the character of the negative viscous friction by fitting predicted oscillations to observed stick-slip.

Experiments grew progressively more sensitive, [Sampson *et al.* 43; Dokos 46; Rabinowicz 51, 56, 58], and evidence mounted indicating that changes in friction do not coincide exactly with changes of mechanism state. That is to say that dynamics exist within the surface processes that determine friction. Using experiments designed to directly determine the properties of break-away (the transition from static to kinetic friction),

Rabinowicz, [51], found that break-away is not instantaneous and proposed a model involving translational distance to account for decreasing friction as motion progressed. In "The Intrinsic Variables affecting the Stick-Slip Process" Rabinowicz, [58], reports an experiment capable of measuring the acceleration of a slider during stick-slip, and observes that the acceleration and deceleration curves are not symmetric. [Rabinowicz 58] is a land mark paper because the two *temporal* phenomena in the stick-slip process are integrated into a model of stick-slip that will correctly predict the range of speeds and structural conditions over which stick-slip will occur. The temporal phenomena are:

1. A connection between the time a junction spends in the stuck condition, dwell time, and the level of static friction (rising static friction);

2. A delay between a change in velocity and the corresponding change in friction (frictional lag).

Dwell Time and Extinguishing Stick-Slip by Increasing Velocity

To understand the role played by rising static friction and frictional lag it is necessary to consider the stages of a stick-slip cycle; this discussion follows [Rabinowicz 58]. In figure 2.14 a pin-on-flat friction machine is sketched. Here the pin is held in place by a spring and the flat moves at a constant velocity. The mechanism is analogous to a servo machine moving with a desired velocity, $\dot{x}_d$, a proportional control gain, k, and damping, b. The discussion assumes moderate damping; extremely large values of damping will influence the qualitative behavior, but moderate values will not [Bell and Burdekin 69].

Under some conditions, a system such as that of figure 2.14 will exhibit stick-slip. The spring force (control action) observed during stick-slip motion is sketched in figure 2.15. During the stuck interval, interval a-b, the force rises at a rate $k\dot{x}_d$. At point b the force reaches $F_{static,infinity}$, the level of static friction when the system has been at rest for considerable time, and slip begins. During interval b-c slip occurs, the exact motion is governed by the mass spring dynamics plus the details of the friction forces. A rapid transit is qualitatively indicated here. At point c the pin is arrested on the flat and the rise at rate $k\dot{x}_d$ begins again, entering a stable limit cycle of points d-e-f. Point d is somewhat lower than point b because the system has only been at rest for dwell time c-d. At point g the velocity $\dot{x}_d$ is increased. The important empirical fact is that as the velocity is increased, the size of the limit cycle, i-j-k, diminishes [Dokos 46; Rabinowicz 58; Kato *et al.* 72]. If the condition at point j were identical to the condition at point d, a decrease

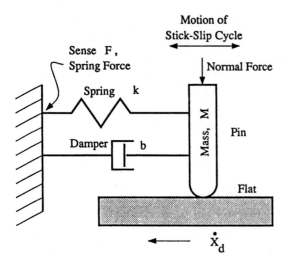

Figure 2.14   Pin on Flat Friction Machine, Schematic.

in the slip distance would not be observed; thus an analysis based on the static plus kinetic friction model will not predict that the limit cycle will decrease. In figure 2.16 the limit cycles d-e-f and i-j-k are shown on a plot of static friction as a function of dwell time. The dwell time is the time during which the surfaces are in fixed contact, the time intervals a-b, c-d, e-f, g-h and i-j in figure 2.15. The static friction increases with dwell time and this accounts for the larger limit cycle at lower velocity. Figure 2.17 is a plot of rising static friction measured directly by Kato *et al.* , [72], who provide a thorough analysis of the processes relating static friction and dwell time. Lubricants A, B, C and D are, respectively, viscous mineral oil, commercial slideway lubricant, castor oil and paraffin oil. Note that in figure 2.16 the time scale is linear, as opposed to logarithmic in figure 2.17. The empirical model of [Kato *et al.* 72], relating static friction and dwell time is:

$$F_s(t) = F_{s\infty} - (F_{s\infty} - F_k) e^{-\gamma t^m} ; \qquad (2.2)$$

where $F_{s\infty}$ is the ultimate static friction; F_k is the kinetic friction at the moment of arrival in the stuck condition; γ and m are empirical parameters. Kato *et al.* , [72], examine conformal contacts and find γ to range from 0.04 to 0.64, and m from 0.36 to 0.67. In this report a a non-conformal contact is examined, γ is found to be 1.66 and m 0.65. A small γ indicates a long rise time and thus resists stick-slip.

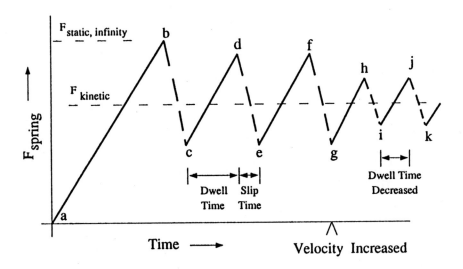

Figure 2.15   Spring Force Profile During Stick-Slip Motion.

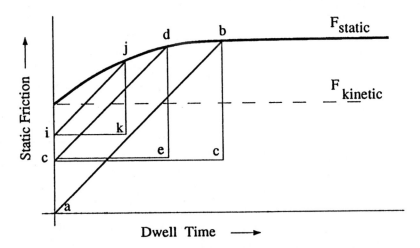

Figure 2.16   Static Friction (Break-Away Force) as a Function of Dwell Time, Schematic; with Stick-Slip Cycle Shown.

Figure 2.18 presents the amplitude of the spring force cycle during stick-slip, shown as a function of machine velocity, $\dot{x}_d$, for several spring stiffnesses, k. The amplitude of the spring force cycle is a decreasing function of velocity until stick-slip is abruptly extinguished. The amplitude is also a decreasing function of stiffness. These data represent several stiffnesses in unlubricated contacts. Brockley *et al.* , [67, 68; Ko and Brockley 70], present

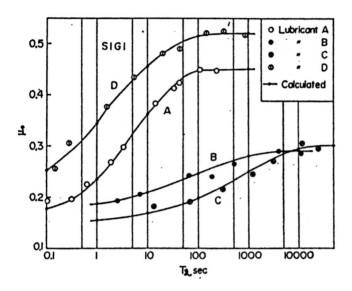

Figure 2.17   Static Friction (Break-Away Force) as a Function of Dwell
Time, Empirical.   (From [Kato *et al.* 72], Courtesy of the
Publisher.)

data observed in an experiment with several levels of damping and Kato
et al. [72] present data collected with various lubricants. The analysis and
data of [Kato *et al.* 72] seem the most germane to typical servo mechanisms
as they incorporate engineering materials and lubricants. All of these data
present the same pattern: slip amplitude as a decreasing function of velocity
up to an abrupt elimination of stick-slip. The process is one of increased
velocity leading to reduced dwell time, which lowers the static friction at
break-away, this further reducing dwell time. At some critical velocity the
dwell time is insufficient to build up destabilizing static friction, and stick-slip
is extinguished. Derjaguin *et al.* [57] present a theoretical treatment that
predicts the critical velocity for termination of stick-slip as a function of
system parameters and the rising static friction. A comparable analysis,
distinguished from that of Derjaguin *et al.* by the incorporation of frictional
lag and the Stribeck effect, is presented in chapter 7. For the controls
engineer, these analyses provide an approach to the question of how slow can
a machine be driven before the onset of stick-slip and on what parameters
does this limit depend.

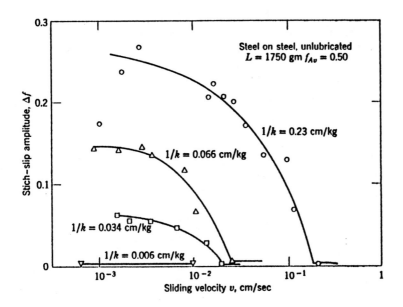

Figure 2.18   Stick-Slip Amplitude as a Function of Velocity, for Several Spring Stiffness.   (From [Rabinowicz 65], Courtesy of the Publisher.)

Frictional Lag and Extinguishing Stick-Slip by Increasing Stiffness

In figure 2.18 one observes that the trial with the stiffest spring did not exhibit stick-slip at any velocity. It is widely observed that stick-slip can be eliminated by stiffening a mechanism [Bell and Burdekin 66, 69; Rabinowicz 65; Armstrong-Helouvry 89]. A stiffness above which there will be no stick-slip is not predicted by a model like that of figure 2.1(a); but increased stiffness is the key to eliminating stick-slip in many mechanical situations [Halling 75].

The Stribeck curve, figure 2.19(a), shows a dependence of friction upon velocity. If there is a change in velocity, one might presume the corresponding change in friction to occur simultaneously, as suggested in figure 2.19(b). In fact there is a lag in the change in friction, as suggested by figure 2.19(c), [Sampson *et al.* 43; Rabinowicz 58, 65; Bell and Burdekin 66, 69; Rice and Ruina 83; Hess and Soom 90]. Returning to the image of partial hydrodynamic lubrication as a water skier with partial dynamic support, if we imagine the water skier half out of the water, his drag will be a decreasing function of velocity. If the tow boat suddenly increases speed, the skiers drag will decrease, but, as in figure 2.19(c), some time will pass before the

new steady state drag is observed. Figure 2.19 is schematic. Experimental
data corresponding to the observation of figure 2.19(c) is presented in figure
2.20.

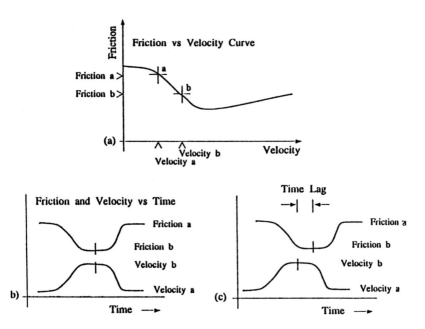

Figure 2.19   Time Relation between a Change in Velocity and the
Corresponding Change in Friction.

Rabinowicz [51] showed that friction level lags a change in system state
with an experiment that related delivered impulse to translation distance in
a sliding contact. He ascribed the lag to a necessary translation distance for
a change in friction, on the scale of surface asperities [Rabinowicz 51, 58,
65]. But there is very considerable evidence that a simple time lag better
describes the effect [Bell and Burdekin 66, 69; Rice and Ruina 83; Hess and
Soom 90]. Bell and Burdekin's [66, 69] data are particularly applicable to
common machine configurations. Figure 2.21 is from [Hess and Soom 90] and
shows friction data for one oscillation of an oscillatory motion that brings
the system into partial fluid lubrication. "μ" in Figures 2.21 and 2.22, as
well as figure 2.17, is the friction coefficient: friction force divided by the
normal load. Note the vertical separation between the friction curves. The
upper friction curve is given during the acceleration away from zero velocity,
and the lower during deceleration. The points of figure 2.21 are experimental

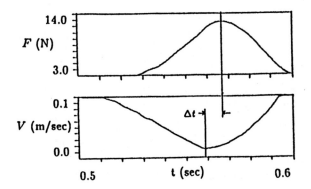

Figure 2.20   Typical Friction-Speed Time Shift; Contact Load = 250 Newtons, Lubricant Viscousity = 0.322 Pa·s, Frequency = 1 Hz. F(N): Friction, Newtons; V(m/sec): Velocity. (From Hess and Soom [90], Courtesy of the Publisher.)

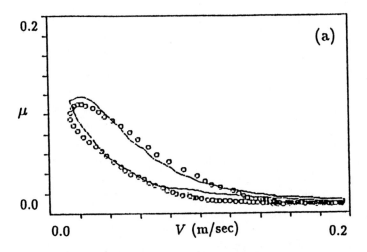

Figure 2.21   Friction as a Function of Velocity; 0: Experimental; —: Theoretical, from Equations (2.3) and (2.4). (From Hess and Soom [90], Courtesy of the Publisher.)

data; the solid line was generated using the friction-velocity model of [Hess and Soom 90], equation (2.4) below, and a pure lag, such that

$$F(t) = F_v\left(\dot{x}\left(t - \tau_L\right)\right) \ . \tag{2.3}$$

where $F_v(\cdot)$ is friction as a function of velocity, equation (2.4), and τ_L is the frictional lag, the time by which a change in friction lags the corresponding change in velocity. Hess and Soom [90] carefully measure the frictional lag and find it to range from 3 to 9 milli-seconds (ms) in a range of load and lubricant combinations; the lag increasing with increasing lubricant viscosity and with increasing contact load. The frictional lag, τ_L, appears to be independent of oscillatory frequency [Hess and Soom 90]. The degree of hysteresis, however, is greatly affected by the oscillatory frequency. When the period of the oscillation is short relative to the frictional lag, the hysteresis, that is the separation between the friction levels during acceleration and deceleration, is greatest. This is illustrated by figure 2.22, also from [Hess and Soom 90]. The data presented were acquired driving their pin-on-disk contact at three different frequencies. Figure 2.22(b) shows the friction curves predicted by their model with frictional lag and should be compared with the experimental data illustrated in figure 2.22(a). Indicative of the progress of tribology, the friction model of Hess and Soom, [90], is to a large degree based on contact and lubricant parameters, only three parameters are fit *a posteriori* to the data.

Evidence for a frictional lag is available from a range of experimental sources: [Sampson *et al.* 43; Rabinowicz 58, 65; Bell and Burdekin 66, 69; Walrath 84; Rice and Ruina 83; Hess and Soom 90]. Tribology is not yet able to offer a theoretically motivated model of the frictional lag, though Xiaolan and Haiqing [87] numerically investigate transient elasto-hydrodynamic lubrication using an analysis that starts with Reynold's equation and Hertzian contact analysis; with this they find a time lag of 3 ms between velocity and friction changes in simulated sliding contact. The physical process giving rise to frictional lag appears to relate to the time required to modify the lubricant film thickness, a process measured by several investigators [Tolstoi 67; Bo and Pavelescu 82; Bell and Burdekin 69]. A period of time required to obtain a new film thickness may be one of several contributing processes, as frictional lag is also observed in dry contacts [Rabinowicz 51]. Whether this process is better modeled by a simple time delay or some other, perhaps linear, formulation is an open question; though experimental data, such as that of [Hess and Soom 90], provides a compelling empirical argument for the simple delay form.

The effect of frictional lag is a pure delay in the onset of the destabilizing drop in friction. From a control standpoint, the frictional lag reduces the destabilizing influence of Stribeck friction. If the time constants of a system are short in relation to the delay in the drop in friction, which is to say that the mechanism (control) is sufficiently stiff, the stick-slip limit cycle will not

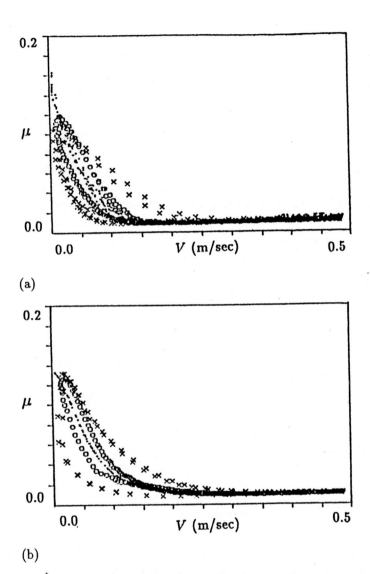

(a)

(b)

Figure 2.22 Friction as a Function of Velocity; For three different Frequencies of Oscillation: ".", 0.1 Hz; "o", 1 Hz; "x", 5Hz. (a): Experimental; (b): Theoretical, from Equations (2.3) and (2.4).(From Hess and Soom [90], Courtesy of the Publisher.)

be stable [Rabinowicz 65]. This is the process whereby increasing stiffness eliminates stick-slip.

Friction as a Function of Steady State Velocity: Variants of the Stribeck Curve

A theoretically motivated mathematical form for the Stribeck friction is not yet available. Figure 2.23 presents several friction-velocity curves. Details of the (f − v) curve depend upon the degree of boundary lubrication and the details of partial fluid lubrication. Curves such as (a) arise when lubricants that provide little or no boundary lubrication are employed. The data of Bell and Burdekin [66, 69] and Hess and Soom [90] indicate such a curve. When boundary lubrication is more effective, the friction is relatively constant up to the velocity at which partial fluid lubrication begins to play a role. Vinogradov *et al.* [66] and Khitrik and Shmakov [87] present data supporting a flat (f − v) curve through the region of boundary lubrication. Fuller [84] cites data contrasting a specific lubricating oil with and without a lubricity additive. The plain oil gives a curve of type (a) ; with the lubricity additive a curve of type (b) is observed ([Fuller 78], figure 14–11; the reference offers considerable discussion of boundary lubrication). One must be careful in discussing friction as a function of steady state velocity. Data collected during velocity transients will exhibit the effects of frictional lag, equation (2.3), and a curve of type (b), figure 2.23, may be observed even if the underlying steady state (f − v) curve is of type (a). Bell and Burdekin, [69], present a thorough analysis of this phenomenon. A curve of type (c) is given by way lubricants [Merchant 46; Wolf 65]. The boundary lubrication provided by the additives to these oils reduces static friction to a level below kinetic friction.

For analysis or simulation it is important to have a mathematical model of the friction-velocity dependence. Hess and Soom [90] employ a model of the form

$$F(\dot{x}) = F_k + \frac{(F_s - F_k)}{1 + (\dot{x}/\dot{x}_s)^2} + F_v\,\dot{x} \qquad (2.4)$$

and show a systematic dependence of $\dot{x}_s$ and F_v on lubricant and loading parameters. Bo and Pavelescu [82] review several models proposed in the literature and adopt and then linearize an exponential model of the form:

$$F(\dot{x}) = F_k + (F_s - F_k)\,e^{-(\dot{x}/\dot{x}_s)^\delta} + F_v\,\dot{x} \; ; \qquad (2.5)$$

where F_s is the level of static friction, F_k is the minimum level of kinetic friction, and $\dot{x}_s$ and δ are empirical parameters. The viscous friction parameter, F_v, is added here; a viscous term was not incorporated by Bo and Pavelescu [82]. In the literature surveyed, Bo and Pavelescu [82] find δ to range from 1/2 to 1. Armstrong-Helouvry [88, 89, 90] employs $\delta = 2$; and

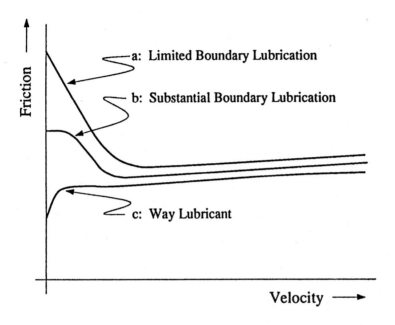

Figure 2.23   Friction as a Function of Steady State Velocity for Various Lubricants; the (f – v) Curve.

the data cited by Fuller [84], observed in a system with an effective boundary lubricant, would suggest δ very large. The exponential model, (2.5), with $\delta = 2$ is a Gaussian model. The Gaussian model is nearly equivalent to the Lorentzian model of Hess and Soom, (2.4), differing only in the tail.

The exponential model, (2.5), is not a strong constraint. By appropriate choice of parameters, curves of types (a), (b) and (c) can be realized. What is needed are data such as that of Hess and Soom [90] over a broad range of engineering materials, conditions and lubricants. For specific lubricant formulations, lubrication engineering firms can provide measures of lubricity and other qualities based on standard industrial tests. Though these tests aren't the equivalent of the experiments of Hess and Soom [90], industrial testing for lubricity is still evolving [Ludema 88].

2.4 Friction Modeling in the Controls Literature

The study of mechanism friction by members of the controls community has been driven by the desire to understand and to compensate for limit cycles observed in mechanism motion, that is stick-slip. The limit cycles themselves produce measurable phenomena: slip distance, period and relative time in stick and slip. These have been the principal data guiding the choice of models for analysis or compensation. Tribology has had little impact on control, even though the friction forces active during the steady motion of lubricated systems are well understood. One reason for the lack of cross pollination is that the dynamic of the friction process around zero velocity is of principal interest in the control of limit cycling mechanisms, but this dynamic has not been of principal interest to tribology, whose dominant concern is wear.

Tustin [47] studied the limit cycling possible in servo-mechanisms with a negative viscous friction characteristic. He does not justify the choice of his friction model, but he is almost certainly affected by the contemporaneous work of Dudley and Swift [49] and others who were finding that stick-slip in mass-spring-dashpot systems could not be explained without a negative viscous friction characteristic. Tustin suggested an exponential decay in friction as a function of velocity, equation (2.5) with $\delta = 1$. Tustin's model is justified because it can explain the macroscopic limit cycle behavior. Tustin's paper is of interest historically because it anticipates describing function analysis by only a few years and achieves much the same result through use of a graphic method of calculation.

Tou and Schulthesis had describing function analysis [Johnson 52] at their disposal and studied the stability of a system with static plus kinetic friction [Tou 53; Tou and Schulthesis 53]. The system includes a proportional-integral-derivative (pid) controller. Limit cycling is predicted; and the authors conclude that it can be eliminated with additional velocity feedback. Tou's work is followed by a large number of extensions which employ the same basic paradigm: selection of a friction model and analysis of limit cycles by the method of describing functions.

Satyendra, [56], studied mechanisms with backlash as well as static plus kinetic plus viscous friction. As did Tou, Satyendra studied stability at zero velocity. Silverberg, [57], strives to separate the describing function of a static plus kinetic plus viscous friction element into frequency dependent and amplitude dependent parts, which greatly aids interpretation.

Shen, [62], studies systems with static plus kinetic friction tracking a ramp position input. The shift in emphasis from the stationary to the slowly moving problem is important. The stationary limit cycling problem, that is, hunting, can always be solved with a dead band in the integral control. It is the explanation of observable limit cycling behavior at low velocities that requires a more sophisticated friction model. Shen and Wang, [64], extend the describing function analysis to third order systems: they propose a control system with gain scheduled derivative feedback and a deadband in the integral control.

Brandenburg *et al.* have studied at length mechanisms with Coulomb friction, backlash and flexibility between the actuator and load [87, 88, 90]. They find that state estimation plays an important role in eliminating the limit cycles introduced by non-linear friction and backlash, and that a reduced order Luenberger observer may provide state estimates adequate to extinguish the limit cycles [Brandenburg and Schafer 1990].

Dither is a well known technique for minimizing the impact of friction, but is not much discussed in the literature. The earliest apparent application of dither is in the control of steam turbines by feedback from a mechanical governor, circa 1890. Because the turbines produced substantially less vibration than older reciprocating machines sticking was observed in the governors and a mechanism was introduced solely to produce vibration [Bennett 79]. [Weaver 59] observes the use of a mechanical eccentric or vibrating member to keep pointing devices in kinetic rather than static friction. [Truxal 55; Oldenberger and Boyer 62; and Atherton 75] also give mention to dither. In the appendix experimental data is presented regarding the application of control dither to a robotic force control experiment. A reduction of a factor of 3 in the rms contact force error is observed with the use of a dither signal that is slightly greater in magnitude than the magnitude of static friction. The early use of dither involved separate vibrating devices, which may reduce the associated wear of actuators and transmissions.

Adaptive control holds the promise of friction compensation without a complete model of the friction dynamic or in cases where friction is slowly varying. Gilbart and Winston, [74], present an ingenious model reference adaptive control system built of analog components, used to control the pointing of a 24 inch, 1400 lb tracking telescope. The authors suggest that the mechanism friction includes a negative viscous friction characteristic, but employ only a kinetic friction term in their adaptive algorithm. Their adaptive algorithm shows a factor of 6 to 10 improvement in rms pointing error relative to a pid controller with no friction compensation.

Walrath, [84], presents a digital adaptive controller designed for an airborne optical pointing and tracking telescope. Walrath's controller identifies and provides feedforward compensation of kinetic friction and the friction transient that occurs with direction reversal. Walrath starts with the Dahl friction model [Dahl 77]. Walrath, however, finds that the data better support a proposed time constant in the friction transient and uses a model form that is related to that of Hess and Soom [90]. Walrath's paper is notable for the careful use of experimental data. His system achieves a factor of 5 reduction in pointing error relative to pid control.

Friction Compensation in Robotics

Compensation for friction took on growing importance in robotics as progress was made toward precise force control. Force control is the governance of contact forces between a manipulator and its environment. Friction becomes important when the error in contact forces is to be held to a level substantially lower than the level of friction in the machine. The challenge is increased by stick-slip: in stiff contact the slip of a stick-slip motion can introduce a considerable uncontrolled change in force. High gain control systems have been proposed, [Wu and Paul 80], as well as feed forward compensation, both of a fixed form [Armstrong-Helouvry 88] or adaptive [Mukerjee and Ballard 85; Canudas de Wit 88, *et al.* 87, 89, 90; Craig 86, 87]. In this report friction is found to be quite repeatable, and, using a friction model with kinetic, viscous and position dependent terms, feedforward compensation is demonstrated accounting for 94% of the friction forces during 3 joint open-loop motions. [Mukerjee and Ballard 85; Canudas de Wit *et al.* 86 and Armstrong-Helouvry 88] find that the kinetic and viscous friction parameters are different in the positive and negative motion directions and a position dependence in the friction. Craig [87] adaptively identifies kinetic and viscous friction parameters, improvement is shown in the performance of the adaptive algorithm relative to a fixed control structure.

[Luh *et al.* 83; Pfeffer *et al.* 89 and Vischer and Khatib 90, 90a] demonstrate the use of output torque sensing, connected with torque rather than velocity control of the actuator, to effect a feedback loop directly around the major friction contributors. The technique, called joint torque control, can produce a factor of 30 reduction in the apparent kinetic friction of the mechanism; a yield sufficient to make joint torque sensing a requirement of a recent NASA request for proposals. Reflecting the fact that the state of the art in force control is advancing, joint torque sensing is available in advanced commercial manipulators [Karlen *et al.* 90]. Note, however, that the authors above report a factor of 30 reduction of apparent static or kinetic friction, whereas the factors of 5 or 6 reported by Gilbart and Winston or Walrath are reductions in RMS output error. Given stability problems at low velocity, these two measures can be compared only with some difficulty.

Kubo *et al.* , [86], study the stability of systems with kinetic friction and pd control. They conclude from phase plane analysis that limit cycling can result and may be dependent on tracking speed, control system parameters and initial conditions. They propose a controller with a fixed kinetic friction feedback that is set conservatively to avoid over-compensation. Experiments show significant improvement in tracking performance when the friction compensation is used. An important aspect of the work is the exploration of the connection between reference trajectory and the effect of non-linear friction. It is observed that the friction may destabilize only some motions and that this is important for empirically tuned controllers. Gogoussis and Donath [87] use phase plane analysis techniques to study the impact of friction in a range of mechanisms.

Townsend and Salisbury, [87], employ describing function analysis to study force control systems with static plus kinetic plus viscous friction and integral control and investigate the stability of system response to various inputs. They conclude that these systems may be stable for some inputs but not for others and point to the implications for empirically tuned controllers.

Repeatability of friction forces is an issue of central importance to the control of machines with friction, yet the literature is virtually silent on the subject. [Rabinowicz, *et al.* 55] analyze the variance of friction forces as a way to probe the processes operating in the sliding interface. Several studies of adaptive friction compensation address the variability of identified friction parameters [Canudas de Wit, Astrom and Braum 87; Craig 86, 87]. Evidence is presented here showing a high repeatability in one servo mechanism. Over all however, the discussion of repeatability could be greatly extended in the experimental literature. Repeatability underlies any model based compensation and is important for considerations of design. The five fold improvement shown by adaptive friction compensation [Gilbart and Winston 74; Walrath 84] and the accurate open-loop motions presented here suggest that friction processes are to a large extent repeatable.

Friction at Zero Velocity

There is a persistent myth in the controls community that friction typically goes to zero at zero velocity. The thinking is that creep will provide a viscous like behavior, albeit with great viscosity. The tribological evidence is for solid bonding. Dahl's model, [Dahl 68], predicts force as a function of displacement and assumes that the asperities remain welded. In the appendix a test for creep is presented during which 90% of the break-away torque was maintained for 18 hours, no motion was observed with a sensitivity of 0.0001 radian. There is little justification for the notion that friction is a continuous function of velocity.

In simulations friction is sometimes represented by a continuous mathematical function, such as $\tanh(\dot{x})$ [Threlfal 78; Rooney and Deravi 82; Ostachowicz 87]. This arises from numerical necessity rather than good tribology, and, while such a model may be quite accurate away from zero velocity, it should be recognized that these simulations will not accurately reflect mechanism behavior in static friction. Others carrying out simulation have implemented discontinuous models [Dupont 90].

For application of the small gain theorem and other results that require a Lipschitz dynamic, a more complete friction model is required. Friction force is a discontinuous function of *steady state* velocity. But through the contribution of asperity compliance and the special dynamics of static friction, friction force is not a discontinuous function of time or true system state. Because of the Dahl effect, static friction is a function of position. As velocity reverses some time is required for the springs of the Dahl effect (see figure 2.7) to windup; and thus some time is required for the transition in friction. Note that for steel parts, the friction transition may occur with a translation of 10^{-5} meters, giving a derivative of friction force that may be an enormous parameter in a small gain calculation.

Special Mechanical Considerations

In practical machines there tend to be many rubbing surfaces - drive elements, seals, rotating electrical contacts, bearings etc. - which contribute to the total friction. In cases where there are several elements contributing at a comparable level, it may be impossible to sort them out and an aggregate friction model may be necessary. In other mechanisms a single interface may be the dominant contributor, as transmission elements often are. The discussion above applies to simple rubbing or rolling friction; in complex machines there may be additional considerations.

One such consideration is different friction in the positive and negative motion directions [Mukerjee and Ballard 85; Canudas *et al.* 87; Armstrong-Helouvry 88]. No explanation for the behavior is available, but it is a sufficient consideration that a standard stick-slip test calls for separate measurements in the left and right directions [Cincinnati Milacron 86]. Another special consideration is normal force. In tribology friction is expressed as a coefficient multiplying the normal force. In mechanisms, the normal force in a load bearing interface may not be a simple function of the system state or actuator command. Studying a mechanism in which preloaded gears dominate the friction, Armstrong-Helouvry [88] finds the friction to be independent of the level of commanded torque. In a lead screw mechanism which was not preloaded, Dupont, [90], finds friction to be proportional to the applied load, and thus employs a coefficient of friction model. Other mechanisms likely fall between these extremes.

Some mechanisms will exhibit position dependent friction [Mukerjee and Ballard 85; Canudas *et al.* 87; Armstrong-Helouvry 88]. Transmissions with spatial inhomogeneities, such as gear drives, will give rise to position dependent friction. As we will see, with accurate friction measurements it is possible to count the gear teeth in the mechanism studied here; and incorporating position-dependent friction in the friction model will substantially increase the accuracy of predicted friction. In part to eliminate position-dependent friction, Townsend [88] studies designs with homogeneous transmissions.

This discussion has focused on friction between lubricated metal pairs. There is discussion in the tribology literature of friction in other machine elements, such as electrical contacts or rotating seals (see, for example, [O'Connor *et al.* 68; Dowson *et al.* 80; Booser 84]. In cases where these elements contribute substantially to the total machine friction, the appropriate model may be quite different from that described here.

2.5 An Integrated Friction Model

This discussion of friction has focused on hard metal parts lubricated by oil or grease. For reasons of machine life and performance, these engineering materials make up many of the machines encountered by controls engineers. When these materials are used, the state of understanding supports a friction model that is comprised of four velocity regimes, two time dependent properties and several mechanism dependent properties.

1. The Four Velocity Regimes:

    I. Static Friction: displacement (not velocity) is proportional to force (see equation (2.1)).

    II. Boundary Lubrication: friction is largely independent of velocity and strongly dependent on lubricant chemistry.

    III. Partial Fluid Lubrication: if static friction is greater than kinetic friction, friction decreases with increasing velocity; this may be destabilizing, but by proper choice of lubricant the instability can be reduced or eliminated.

    IV. Full Fluid Lubrication: friction is a function of velocity, a viscous plus kinetic friction model may model the friction quite accurately. (Regimes II-IV, see equation (2.5)).

2. The Two Time Dependent Properties:

    I. Rising static friction with increasing dwell time (see equation (2.2)).

II. Frictional lag: in partial fluid lubrication friction is dependent upon velocity and load; a change in friction will lag changes in velocity or load (see equation (2.3)).

3. The Several Mechanism Dependent Properties:

   I. The friction parameters may be different in the different directions of motion.

   II. In mechanisms with non-homogeneous transmissions, friction may be position dependent.

   III. In mechanisms with little preloading, friction may be dependent upon instantaneous torque or load; in mechanisms with preloaded transmission elements and roller bearings, friction may be to a large degree independent of torque or load.

Tribology, from its origins, has been concerned with wear and the loss of energy to friction. Controls is concerned with dynamics. In the early days of tribology, the physical processes of friction were under investigation and dynamics was often one means to observe these processes. As the basic mechanisms of friction became better understood, however, these simple experiments gave way to the development of more rigorous theory focusing on surface physics and chemistry, the hydrodynamics of lubrication, and on the phenomena of wear. Thus the controls engineer finds relevant studies in the tribology literature of the forties and fifties, and then a long silence on the dynamics of a machine with friction. A reunion is coming. Using refined numerical methods and more powerful computers, several authors discussed above probe the transient dynamics of fluid lubrication. Partial fluid lubrication is coming under investigation; and even transient partial fluid lubrication, the essence of stick-slip, is considered.

Evidence suggests that in servo mechanisms friction can be highly repeatable. Repeatability, coupled with knowledge of the relevant model structure, will permit experimental identification of the friction model describing a particular mechanism. The availability of an accurate friction model will permit analytic prediction of performance, correct calculation of time or energy optimal control, correct decoupling of multi-degree-of-freedom mechanisms and the design of friction compensation. In cases where friction introduces stick-slip, lubricant choice should be carefully considered and lubricity additives evaluated. Where possible, stick-slip should be eliminated by proper choice of lubricant. When stick-slip remains, an accurate friction model will aid in the determination of mechanism and control performance required to achieve smooth motion.

Chapter 3

Experiment Design

The experiments reported here have been carried out with a PUMA 560 arm controlled with the NYMPH computer [Chen *et. al.* 86]. Two procedures for measuring friction have been used:

1. Measuring acceleration and subtracting computed inertial torques from known motor torques;
2. Measuring the minimum torque necessary to initiate motion.

Acceleration Measurement

Measuring acceleration is a substantial challenge. Using a two-sided Kalman smoother, acceleration can be estimated from recorded position data; however if noise rejection is set to an adequate level, bandwidth is quite poor. In trials with PUMA position data, acceptable noise rejection was determined to be 18 dB reduction of white quantization noise, which gave an RMS acceleration deviation of 0.3 rad/sec^2. This level of noise rejection results in an effective bandwidth of roughly 2 hertz, a severe experimental limitation. The Systron-Donner corporation generously lent a model 8160 rotational accelerometer to this project. This extraordinarily sensitive instrument is specified to have a threshold and resolution of 0.005% of full scale; in the case of this instrument, that is 1 milli-radian per second squared. The response of the instrument is flat to a second order roll-off at 40 hertz. An accurate acceleration measurement has been key to the measurement of friction during motion.

Velocity Estimation

In most of the experiments presented here, velocity was estimated off line using a two pass Kalman smoother running on position and torque data. This procedure gives a reliable, unbiased estimate of velocity, and has a bandwidth of roughly 10 hertz. For the low speed trials reported in chapter

6, a real time velocity estimate was required: a digital cross-over filter was implemented that had the high frequency response of the integrated acceleration signal and the low frequency response of the differentiated position measurement. The cross over frequency was 10 radians per second.

Contact Force Sensing

Measurements of contact force were made with the Stanford force sensing fingers (see [Khatib and Burdick 86]). The force finger sensor consists of a three axis load cell with a stiffness of 60,000 Newtons per meter. The sensitivity is roughly 0.01 Newtons, or 0.04 ounces.

Torque Control

The Unimate controller employs high gain current amplifiers to drive the motors. During operational tests the current was observed to slew to a new command value in less than 500 μsec and hold that value to within 0.5%. The motor model used here, the standard dc motor model, provides that torque is proportional to current and independent of velocity; a multiplicative torque ripple of a few percent may be included. The PUMA motors in particular, which are Magnetic Technologies model 3069-381-016, are specified to have a torque constant of .261 Newton-meters (N-m) per Amp, and an average to peak ripple of 4% at 25 cycles per revolution. The maximum available motor torque is 1.70 N-m; reflected through the transmission, this gives a maximum torque of 100 N-m at joint 1.

Break-Away Torque

The break-away torque was measured by stopping the arm, setting the torque to zero (or to the gravity compensation value, if any) and ramping up the torque in 0.2 Newton-meter steps at a rate of 40 steps per second. Break-away was established when motion (one or more shaft encoder pulses) was observed during two consecutive steps. The torque applied during the first interval with motion was taken to be the break-away torque. Highly variable "wind-up" was observed in the time prior to steady motion, and labeling the first shaft encoder pulse to arrive as the break-away event led to non-repeatable results. The step size and rate of this experiment were chosen to maximize repeatability.

Linear Parameter Estimation

When parameters to be identified appear linearly in a model, as will be the case in the main of these experiments, solving the normal equations produces an estimate of the parameters which minimizes the squared error. The estimation is constructed in the following way:

If the model may be written

$$\tau(k) = \phi^{\mathrm{T}}(k)\,\theta^* \tag{3.1}$$

where $\tau(k)$ is the torque applied at time step k;
$\qquad\quad\phi$ is the regressor vector, it contains functions of the manipulator state
$\qquad\quad\theta$ is the parameter vector, it contains fixed parameters;
$\qquad\quad\theta^*$ is the vector of true parameter values;
$\qquad\quad\widehat{\theta}$ is the estimated parameter vector;
$\qquad\quad k$ is the time step in the process;

then the normal equations may be constructed by writing

$$\mathbf{T} = \begin{bmatrix} \tau(0) \\ \vdots \\ \tau(K) \end{bmatrix} \qquad \Phi = \begin{bmatrix} \phi^{\mathrm{T}}(0) \\ \vdots \\ \phi^{\mathrm{T}}(K) \end{bmatrix}$$

and estimating the parameters according to:

$$\widehat{\theta} = \left[\Phi^{\mathrm{T}} W \Phi\right]^{-1} \Phi^{\mathrm{T}} W \mathbf{T} \tag{3.2}$$

where W is the covariance matrix of the data vector $\mathbf{T}$.

When the normal equations are used to estimate friction parameters, ϕ will commonly comprise the joint acceleration, velocity and sign(velocity):

$$\phi = \begin{bmatrix} \ddot{q} \\ \dot{q} \\ sign(\dot{q}) \end{bmatrix} \qquad \theta = \begin{bmatrix} mass \\ viscous\ friction \\ kinetic\ friction \end{bmatrix}$$

So that

$$\tau(k) = mass * \ddot{q}(k) + (viscous\ friction) * \dot{q}(k)$$
$$+ (kinetic\ friction) * sign(\dot{q}(k)).$$

Once the parameters are computed, it is possible to compute the residual error according to:

$$\widetilde{\mathbf{T}} = \mathbf{T} - \Phi\,\widehat{\theta}\ ; \tag{3.3}$$

where $\widetilde{\mathbf{T}}$ is torque not predicted by the model.

And to compute the weighted residual variance according to:

$$\sigma^2 = \widetilde{\mathbf{T}}^{\mathrm{T}} W \widetilde{\mathbf{T}} \ .$$

Non-Linear Estimation

In chapter 6 velocity parameters which appear in a nonlinear friction model are estimated. This was done by a mesh search: the weighted residual variance was evaluated at all points on a mesh and the minimum selected. The method has the liability of high computational cost. But in each case where it was applied, local minima were observed. When the fit need only be done once, searches with compute times of days are not inpractical.

Low Pass Filtering

In some cases below, the data presented have been low pass filtered. To avoid the phase lag associated with causal filter implementations, this has been done off line using convolution with a Gaussian curve. Where a cut-off frequency is cited, this is the width to half maximum of the filter frequency response.

Chapter 4

Repeatability

Perhaps the most fundamental issue in an effort to model any process is repeatability. Whatever hope exists of capturing the process in a predictive model is predicated upon repeatability. Tribologists have studied the variance of friction forces in carefully controlled situations [Rabinowicz, et. al. 55; Rabinowicz 56]. But it is not straightforward to extrapolate from these data to complex mechanisms. Repeatability may be inferred from the successful compensation of friction demonstrated by Gilbert and Winston, [74], and Walrath [84]. Walrath in particular presents correlation data that suggest a high degree of repeatability. But neither of these investigators explicitly addresses repeatability.

Here I undertake to measure repeatability in the simplest possible way: by playing out a pre-determined sequence of torques and observing the motion. A stiff PID controller, the standard Unimate controller, is used to pre-position the arm. The industrial controller is able to attain a desired position to within ± 0.001 radians. Using precomputed torques, as opposed to closed-loop control, ensures that the applied torques are the same from trial to trial. A typical result is presented in figure 4.1 where the velocity profiles of three successive motions are shown.

Prior to collecting these data the arm was "warmed up" by a minute of motion throughout the workspace. The effect and importance of warming up the mechanism are discussed in section 5.1. After warm up the open-loop trails were run using the torque sequence of figure 4.2. The three velocity profiles of figure 4.1 reveal considerable structure: the arm repeatedly accelerates and decelerates in a roughly sinusoidal pattern. If the velocities are plotted as a function of position, the correspondence becomes more dramatic, as shown in figure 4.3.

Using accelerometer data it is possible to determine the mass, viscous friction and kinetic friction observed during the motion by solving the normal

47

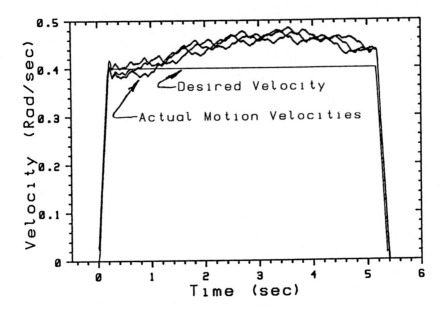

Figure 4.1   Velocity Resulting from the Application of a Constant Torque
to Joint One of a PUMA 560 Robot.

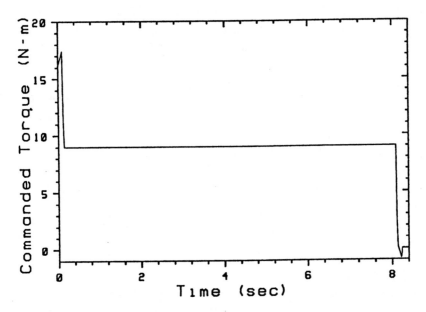

Figure 4.2   Torque Applied During Each of the Three Motions Depicted in
Figures 4.1.

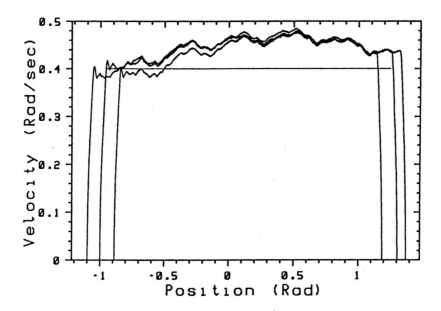

Figure 4.3 Velocity Profiles of Figure 4.1 Plotted Against Position; Desired Velocity Shown.

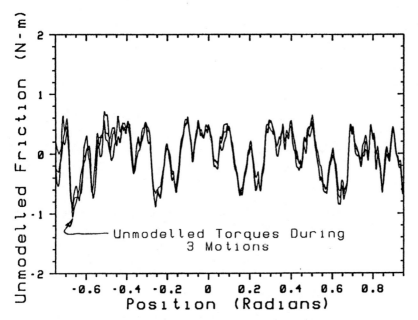

Figure 4.4 Residual Torques of a Mass plus Viscous plus Kinetic Friction Model, Plotted Against Position. (Low Pass Filtered to 20 Hz.)

equation, (3.2). Equation (3.3) then gives the residual torques, the torques which are unexplained by the model. The residual torques for the motions of figure 4.3 are plotted in figure 4.4, here they are shown as a function of position.

By sampling laterally across several data sets, it is possible to estimate the variance in friction torque. To produce figure 4.5 the residual friction torques were determined as they were for figure 4.4. At each time step the residuals of five motions (including the three of figure 4.4) were averaged and the variance determined. Figure 4.5 is a plot of the average across these five moves; note that large residual torques occur at the beginning and end of the trajectory. Figure 4.6 is more interesting; it is a plot of the standard deviation of the data in each of the bins made by the averaging process which gave figure 4.5. The standard deviation is remarkably flat.

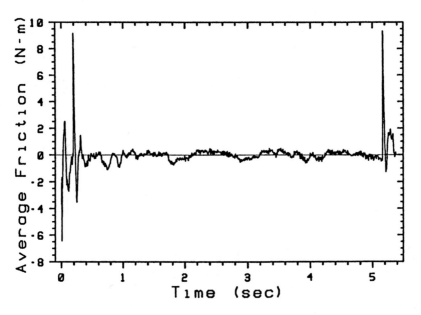

Figure 4.5(a)   Average of the Residual Torque During Five Repetitions of the Motion of Figure 4.1. The Data have been Binned by Time. Scale Chosen to Show Peaks.

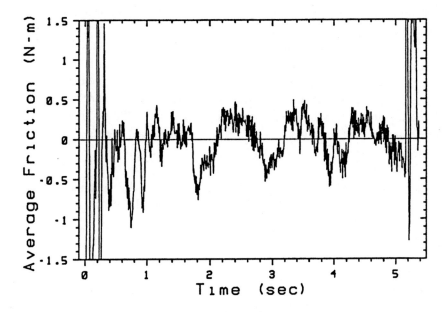

Figure 4.5(b) Average of the Residual Torque During Five Repetitions of the Motion of Figure 4.1. The Data have been Binned by Time. Scale Chosen to Show Detail and Match Figure 4.6.

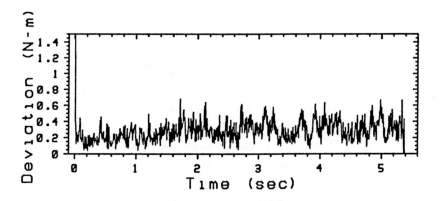

Figure 4.6 Standard Deviation of the Residual Torque During Five Repetitions of the Motion of Figure 4.1. The Data have been Binned by Time.

Figure 4.4 shows a high degree of position dependence in the torque error. This dependence is confirmed in figure 4.7, which is a plot of average and deviation in the residual friction of the five trials, as in figures 4.5

and 4.6, but here they are averaged and plotted as a function of position. During the cruise portion of the trajectory, the deviation of figure 4.7 is quite small. In table 4.1 the magnitude of the friction torque, the residuals and residual deviations are presented. The deviation of the residuals is of greatest interest: the deviation is a measure of the non-repeatable component of the friction process. The degree of repeatability seen here is remarkably great, the deviation in friction between identical trails is only 1% as great as the magnitude of the friction itself: 0.107 [N-m] of Deviation versus 10.4 [N-m] of Friction.

Table 4.1   Magnitude of Friction Torque Captured by a Kinetic plus Viscous Model and of the Residuals and the Deviations of the Residuals.

Friction Signal	Magnitude (N-m)
Modeled Friction	10.407
RMS Residual, Sampled According to Time Along the Trajectory	0.282
Mean Standard Deviation of the RMS Residual, Time Sampled	0.284
RMS Residual, Sampled According to Position Along the Trajectory	0.379
Mean Standard Deviation of the RMS Residual, Position Sampled	0.107

The reduction of deviation arising from correlating the friction data with position, from $\sigma = 0.284$ to $\sigma = 0.107$ corresponds to a Fisher statistic of $F_{(1000,500)} = 12.1$. Use of the Fisher statistic is described in greater detail in section 6.1. Let it suffice here to say that the F statistic can be used to test whether an improvement in model accuracy is achieved by chance or by actually explaining an underlying systematic process. In this case an $F_{(1000,500)}$ of 12.1 yields a confidence of 99.99999999% that a systematic correspondence between position and friction exists.

In this chapter a fundamental aspect of friction has been addressed: repeatability. Surprising repeatability is observed. Study of the friction force that is not explained by a kinetic plus viscous friction model shows a substantial component which, with very high confidence, is systematically correlated with position. Measuring this position-dependent friction will be the subject of the next chapter.

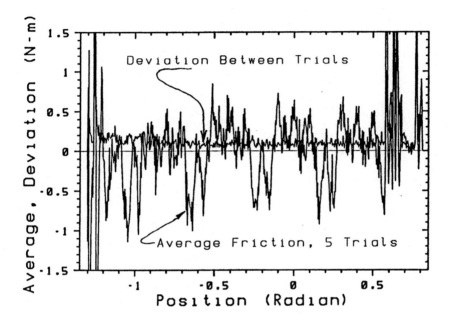

Figure 4.7   Average and Deviation of the Residual Torque During Five Repetitions of the motion of Figure 4.1. The Data have been binned by Position.

Chapter 5

Break-Away Experiments

The break-away experiments consist of allowing the arm to come to rest (static friction) and ramping up torque, measuring the torque required to initiate motion. The beauty of the break-away experiment is its ability to sample static friction at extremely high spatial frequency. The break-away experiment does not suffer from problems of mechanical compliance, phase lag and instrument bandwidth that are associated with motion experiments. The data are useful because they correlate well with the position-dependent disturbance observed in the motion data. An example of the alignment between break-away data and the unexplained friction in the motion data is shown in figure 5.1, where the solid trace is the error torque experienced during a motion, similar to that presented in figure 4.4. The dashed trace is taken from break-away data, suitably scaled and offset to align with the motion data. Both traces have been low pass filtered to 20 cycles per radian. As an aside, the break-away data were collected on June 6, 1987, and the motion data on January 12, 1988, six months and a thousand machine operating hours later.

5.1 Experimental Issues in Measuring Break-Away Torque

Four aspects of this experiment were tuned to achieve maximum repeatability: a warm up exercise was developed, fine sampling was used, the torque rate was tuned, and the definition of break-away selected to maximize repeatability. Early in this experimental work it was observed that after an extended period of inactivity (over night) the friction would start high and diminish quickly to a steady value. A series of empirical trials showed the motion friction to come to its steady value after a minute of moderate velocity activity spanning the work space. The motion friction would remain at its steady value through tens of minutes of inactivity. Like the motion friction, the static friction was observed to increase with over night inactivity.

55

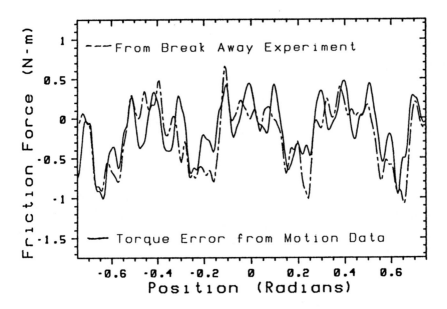

Figure 5.1. The Alignment of Break-Away Torque Data and Motion Friction Data.

Unlike the motion friction, the static friction was also observed to increase with tens of minutes of inactivity. The rising static friction is perhaps a consequence of the interaction of static friction and dwell time, see section 2.3. To improve repeatability, the longer time process was addressed with a vigorous motion at the beginning of each session. While sampling the static friction, a shorter back and forth motion, spanning the space to be sampled during the coming interval, was carried out about once per minute.

The break-away data were observed to have features at extremely high spatial frequencies. The initial sampling density was inadequate and problems arose from aliasing and bias introduced by the sample location selection. To collect break-away data at a spatial resolution higher than that achievable by the standard industrial controller, an open-loop type sample location selection was used. The Unimate controller can achieve commanded positions to within ± 0.001 radians, whereas the shaft encoder has a resolution ten times finer. Rather than attempting to control the motion to a desired sample point, a sample was simply collected at the location of the arm. Each sample moves the arm forward a distance of several bins. After many (possibly ten) sweeps over each region nearly all of the locations will have been sampled. Experience showed that some locations were unreachable, either by feedback control or by chance !!   I believe that the steep friction gradient surrounding these points made them unreachable. Quadratic interpolation was later used to estimate the friction at these locations. For building the tables of position-dependent friction, 2,000 bins per radian were used. The highest prominent spatial frequency observed is 355 cycles per radian; thus 2,000 samples per radian provides nearly a factor of three head room above the Nyquist limit.

The break-away experiments required considerable time (60,000 data points ×2 directions ×3 joints ×2 seconds per point = 200 hours). It was therefore important to maximize the rate of the experiment. As roughly 80% of the experimental time was spent ramping up the torque, the size of torque step and step rate were critical experimental issues. The results of several trials are presented in table 5.1; 0.2 [N-m] per step and 40 steps per second was chosen as a good compromise between repeatability and speed. The result of this effort toward achieving good repeatability in the break-away measurements is shown in figure 5.2. The three lines of the figure are, for each bin, the mean and plus and minus one standard deviation. The mean static friction of figure 5.2 is 8.43 N-m, the average of the bin-wise standard deviations is .237 N-m, or 3% of the mean.

Table 5.1 Measurement of Deviation in the Break-Away Torque as a Function of Experimental Parameters.

Sample Rate	Torque Step Size	Mean Break Away Torque	Standard Deviation
(Hz)	(N-m)	(N-M)	(N-m)
40	.2	8.51	0.165
40	.1	8.63	0.151
40	.4	9.00	0.330
80	.2	9.10	0.220
80	.1	8.86	0.124

Initially the break-away torque was taken to be that applied when the first shaft encoder pulse was observed. This led to highly non-repeatable results. Trials were made requiring two, three and four shaft encoder pulses to certify motion and determine the break-away torque, but with little improvement in repeatability. Imposing what is in essence a velocity requirement, that shaft encoder pulses come in two consecutive intervals, markedly improved the repeatability and was adopted as the condition to certify break-away. The combination of 0.1 N-m steps at 80 Hz, which is suggested by table 5.1, was rejected because of the velocity and thus acceleration required to achieve two shaft encoder pulses in consecutive

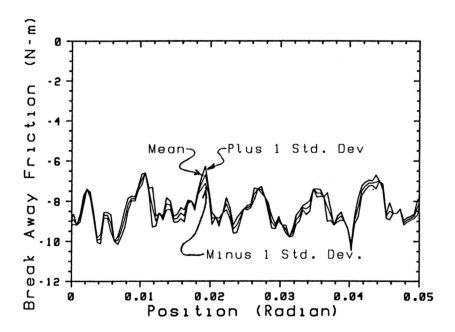

Figure 5.2 A Portion of the Break-Away Data on an Expanded Position
Scale. The Mean and Plus and Minus One Standard Deviation
Curves are Shown.

sample intervals. The lack of repeatability in the observation of the initial
pulse has been connected to the Dahl effect (see chapter 6.4).

5.2 Building the Compensation Table

The break-away experiment provides an excellent means to measure
position-dependent friction and develop a friction compensation lookup
table. To capture phenomena at spatial frequencies as high as 355 cycles
per radian a sampling frequency of 2,000 bins per radian was selected. To
prepare a lookup table for 5 radians of joint motion, 10,000 bins were needed.
Collecting 60,000 data points allowed averaging and improved accuracy.

Examination of the curves of figure 5.1 suggests that there should be
a few dominant spatial frequencies. The major periodicity makes 3 cycles
across the figure, and a less prominent periodicity appears at about 5 cycles
per cycle of the major periodicity. A spatial FFT of the break-away data
fails, however, to reveal these periodicities, as shown in figure 5.3. Figure 5.3
is a spatial FFT of the joint 1 break-away data set, sampled at 2,000 samples
per radian, low pass filtered to 100 cycles per radian and pre-processed with
a Blackman-Harris window. The result is a flat transform, not showing the
dominant periodicities that the eye sees in figure 5.1. The major periodicity

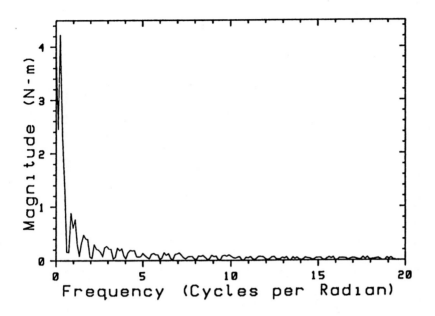

Figure 5.3 Spatial FFT of the Break-Away Data for Joint 1 of a PUMA 560 Arm. Note the Lack of Periodicities Apparent in Figure 5.1 or Figure 5.7.

of figure 5.1 makes 2.8 cycles per radian or 15 cycles per revolution of the joint. This is the rotation frequency of the intermediate gear of joint one, suggesting that periodic friction phenomenon is related to a non-uniformity of the intermediate gear. The higher frequency apparent in figure 5.1 occurs once per revolution of the motor, suggesting a non-uniformity in the motor or motor gear.

To correct for position-dependent friction a lookup table was constructed using data from the break-away experiment. The velocity profiles of three open-loop motions are shown in figure 5.4. Applying the position-dependent compensation from the lookup table reduced the variance of the velocity during the motion. It was observed that the table compensation was most effective at lower spatial frequencies, the variance in velocity and acceleration was measured with compensation tables applied that were low pass filtered to dc (0 Hz), 0.8, 4.0, 20, 100, and 500 cycles per radian (the raw data were binned at 2,000 samples per radian). The deviation (square root variance) in velocity measured along the motion of figure 5.4 is shown as a function of

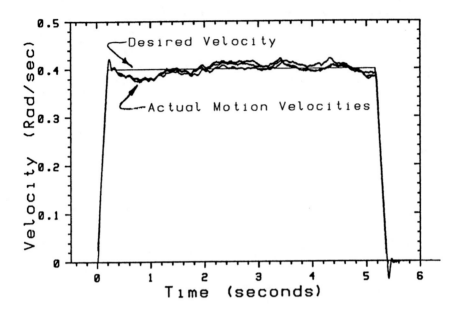

Figure 5.4   Velocity Profile Recorded During an *Open-Loop* Motion of the
Arm with Position-Dependent Friction Compensation Applied;
contrast with figure 4.1, for which no Position-Dependent
Friction Compensation was Applied.

the table bandwidth in figure 5.5. In figure 5.6 the deviation in acceleration
is shown.

The best reduction in velocity deviation is roughly a factor of 2. There
is no apparent reduction in acceleration deviation. The table lookup scheme
predicted the position one and a half sample times ahead, to try to select the
position corresponding the the middle of the sample interval during which
the torque would be applied. The break-away data were low pass filtered to
20 cycles per radian and then compressed into a lookup table with a sampling
frequency of 200 bins per radian. At the moderate arm velocity of 1 radian
per second, the arm would cross only one bin during one 5 milli-second cycle
of the control system. No attempt was made to average the compensation
from several bins according to the velocity of the arm.   After filtering,
an offset was added to make the lookup table zero-mean: kinetic friction
compensation is provided as a separate term. The resulting compensation
torque table is shown graphically in figure 5.7, this compensation was used
to generate the motion of figure 5.4, the velocity profiles of which should

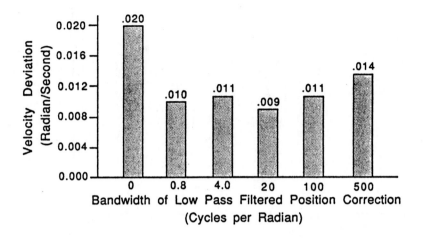

Figure 5.5 RMS Deviation in Velocity as a Function of the Spatial Bandwidth of the Position-Dependent Friction Compensation Table.

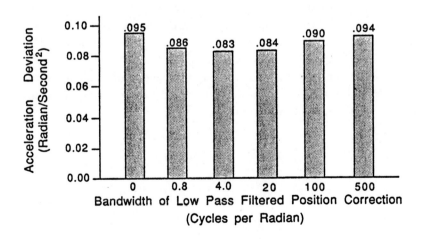

Figure 5.6 RMS Deviation in Acceleration as a Function of the Spatial Bandwidth of the Position-Dependent Friction Compensation Table.

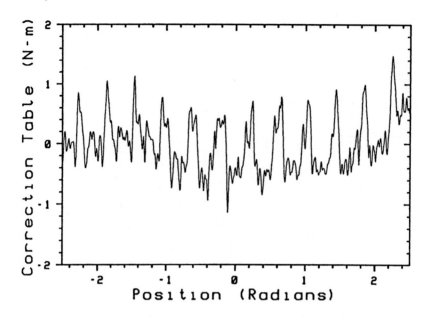

Figure 5.7   Lookup Table Compensation Torques Applied at Joint One of
the PUMA 560.

be contrasted with those of figure 4.1, for which no compensation table was
used.

In this chapter an experiment that can accurately measure friction as
a function of position is presented. Again the repeatability is striking, this
time the repeatability of the torque required to achieve break-away. The
importance of modeling the position-dependence in the friction for control
will depend upon the application. But the importance of modeling this effect
for other experimental work is tremendous; the model of position-dependent
friction will be used many times in what is yet to come.

Chapter 6

Friction as a Function of Velocity
(Negative Viscous Friction Revealed)

The control engineer often assumes a viscous friction model. To describe dry rubbing the tribologist rarely includes any velocity dependence at all, and when he does it is as likely to be negative as positive. Based on the experimental work presented here, the kinetic plus viscous friction model seems very accurate at velocities above a minimum velocity. Below the this velocity, decreasing friction with increasing velocity, or negative viscous friction, is observed.

The relationship between friction and velocity has been measured in three ways:

1. Open-loop, constant torque motions;
2. Closed, stiff velocity loop, constant velocity motions;
3. In contact, an open-loop torque ramp against a compliant surface.

Each of these experiments offers an advantage: the open-loop gliding motions provide a test bed for measuring the repeatability of the friction forces - a basic objective of this research - and to test for the dependence of motion friction upon position and load. The closed-loop control provides a means to penetrate the unstable, low velocity regime. And the compliant motions - the most illuminating experiment of this research - provide a means of mapping the entire regime of unstable velocities.

The PUMA arm is pictured in figure 6.1, a photo taken during the three joint open-loop motions described in section 8.2. Joint one, the proximal joint, provides rotation about a vertical axis coincident with the supporting column. Joint one was chosen to be the experimental testbed because it has the greatest inertial, which helps to bring the the motion phenomena within the sensing bandwidth, and because the vertical axis does not bear gravity load. Figure 6.2 is a plot of friction as a function of velocity, measured on joint one of the PUMA arm. These data were collected by

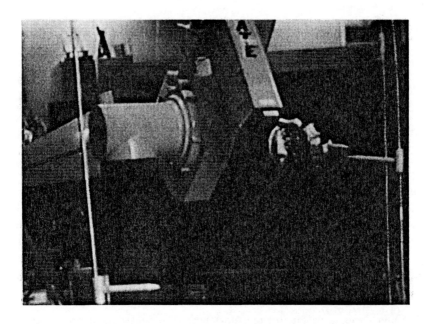

Figure 6.1   The PUMA Arm with Indicators Used for Open-Loop Motion
                 Trials.

applying a constant torque and observing the average resulting velocity. The
position-dependent friction was compensated by a zero-mean table lookup
torque, as described in section 5.2.

6.1 Analysis of Variance in the Motion Friction Data

Figure 6.2 is based on five measurements of velocity, at each of four
positions, at each of four constant torque levels, in the positive and negative
directions: a total of 160 measurements. This data set is sufficient to
allow the use of standard analysis of variance techniques to test whether
separate positive and negative direction parameters should be used, and to
test whether the motion friction is dependent upon position. The analysis of
variance (ANOVA) is a technique that allows the variance explained by the
extension of a model to be compared with the residual variance. When linear
estimation is used, adding parameters to a model will nearly always result in
an improvement in the fit of the model to the data. Using ANOVA and the

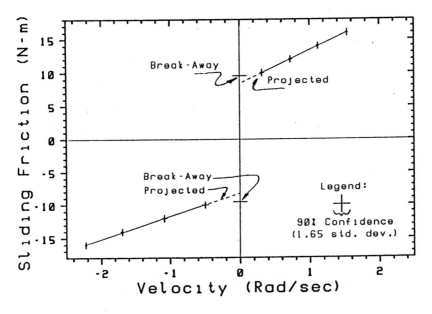

Negative Direction
Viscous 3.45 N-m/rad/sec
Kinetic 8.26 N-m

Positive Direction
Viscous 4.94 N-m/rad/sec
Kinetic 8.43 N-m

Figure 6.2   Friction Torque as a Function of Velocity. (Note that the Kinetic and Viscous Parameters are Different for Positive and Negative Direction Rotations.)

Fisher statistic, F, a standard test of significance, it is possible to determine whether the improvement given by adding model parameters is significantly better than chance and thus justified. The test consists of comparing the improvement achieved per degree of freedom added to the model, the MST, with the residual variance per degree freedom, MSE. If the extra parameters have nothing whatever to do with the source of variance (which may simply be random), one would expect the F statistic, $F = MST/MSE$, to be nearly one. A large F statistic indicates that considerable variance is explained by the extension of the model and that the increased model size is justified. The F statistic required to accept the hypothesis is a function of the confidence level required and the number of degrees of freedom involved. The statistical test, tempered by physical modeling, provides a means of testing the merit of proposed models.

The hypothesises to be tested are:

Hypothesis 1:   The model should include separate kinetic and viscous friction parameters for the positive and negative rotation directions.

Hypothesis 2:   There is position dependence in the kinetic and viscous parameters, and separate parameters should be used for each of the four starting points used. This would be a coupling between position and motion friction in addition to that corrected by the table lookup compensation.

The variance measures are defined to be:

MST: Mean squared error explained by the use of a larger model.

MSE: Mean squared error still unexplained, even with the use of a larger model.

Table 6.1   ANOVA Testing of Hypothesis 1 and 2. (Squared errors are in units of $[(\text{N-m})^2]$.) ($F_{n,m}$ indicates the measure with n additional parameters and m remaining degrees of freedom.)

	MST	MSE	F	Threshold F, 95% Confidence	Accept Hypothesis?
Hypothesis 1	13.209	0.00605	2182.0	2.15 $(F_{2,28})$	Yes
Hypothesis 2					
Positive Direction	0.00475	0.00295	1.61	9.03 $(F_{6,8})$	No
Negative Direction	0.00828	0.00847	0.98	9.03 $(F_{6,8})$	No

The use of positive and negative direction kinetic and viscous parameters is strongly indicated (Hypothesis 1). The position dependence of the kinetic and viscous parameters (Hypothesis 2) is not supported, the F value of 1.61 is little better than chance and the F value of 0.98 is worse than chance.

6.2 Friction at Low Velocities

Stick-slip oscillation during low speed motion has been observed by many investigators and a sizable controls literature exists studying the contribution of static friction to this oscillation (see section 2.4). In the literature outlined in section 2.4, describing function or phase plane analysis is applied to a static friction model, such as that of figure 2.1(b); the outcome is oscillations when integral control is used. None of the work based on a simple sticktion model can explain oscillation without an integral control term. Of the authors in the controls literature addressing stick-slip, it is surprising that none recommends restructuring control to omit the integral term.

During unrelated experimental work at Stanford [Khatib and Burdick 86] unstable motions were observed even when no integral control term was used. The phenomenon is explained by negative viscous friction. Tustin [47] was the first to make use of a model with negative viscous friction in the analysis of feedback control. Tustin's model is an example of (2.5) with $\delta = 1$:

$$F(\dot{x}) = F_k \operatorname{sgn}(\dot{x}) + F_v \dot{x} + F_s e^{(-\dot{x}/\dot{x}_s)} \qquad (6.1)$$

where $F(\dot{x})$ is the friction as a function of velocity;
        F_k is the kinetic friction;
        F_v is the viscous friction parameter;
        F_s is the difference between static friction and kinetic friction;
        $\dot{x}$ is the motion velocity;
        $\dot{x}_s$ is a constant with units of velocity giving the characteristic velocity at which the system transits to kinetic friction.

Tustin predicts oscillations at low speed, even in the absence of integral control. Tustin provides no experimental evidence, either direct or cited, to support his choice of model structure; none-the-less, Tustin's contribution stands out: it explains an observable phenomenon neglected by a host of authors.

To measure the motion friction at low velocities, a stiff velocity control loop was implemented and the average torque required to sustain steady motion was measured. With a sample rate of 200 Hertz and using the first difference of position as an estimate of the velocity, it was not possible to implement a velocity error gain greater than 50 Newton-meters (N-m) per rad/sec. This gain permitted a minimum velocity of 0.15 rad/sec before the onset of stick-slip. The standard industrial controller also exhibits a minimum velocity of 0.15 rad/sec before the onset of stick-slip. By integrating the accelerometer signal, the velocity error gain could be increased to 90 N-m per rad/sec, which allowed testing motions as slow as 0.015 rad/sec. A velocity of 0.012 rad/sec was achieved with direct accelerometer feedback. Note that using only the position measurement and a simple control structure (the standard industrial implementation), the apparent minimum velocity of 0.15 rad/sec is 7% of the maximum velocity achievable at the joint.

The measurements made with a stiff velocity loop are shown in figure 6.3, a downward bend in the low velocities is clearly evident. Five samples were taken at each velocity, from this the deviation was estimated giving the 90% confidence intervals shown.

Fitting Tustin's model to the variance weighted data gives $\dot{x}_s = 0.019$ rad/sec. The friction is given by:

$$F(\dot{x}) = 8.43\,\text{sgn}(\dot{x}) + 4.94\,\dot{x} + 1.13e^{(-\dot{x}/0.019)} \quad [\text{N} - \text{m}]. \qquad (6.2)$$

(Viscous friction term taken from the fit to high velocity data).

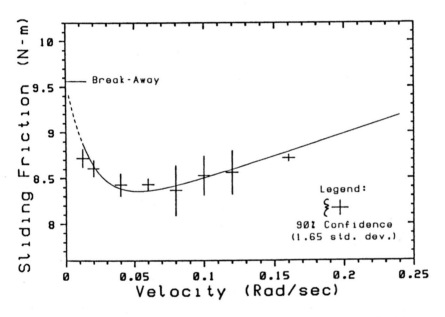

Figure 6.3   Friction as a Function of Velocity at Low Velocities, curve given by Tustin's Exponential Model.

6.3 Friction During Compliant Motion

The data collected by closed-loop motions show clear evidence of an upward turn in the friction curve at low but significant velocities. However the model and experiment are unsatisfactory in two respects: the model does not fit the data well, the lowest velocity data point showing a 2 standard deviation excursion from the best fit model; and no measurements

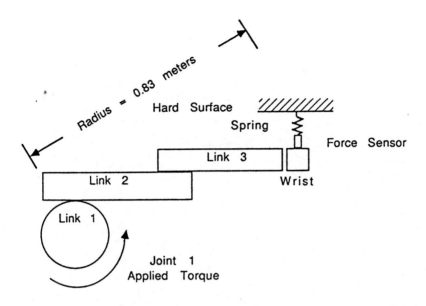

Figure 6.4 Apparatus of the Compliant Motion and Force Control Experiments.

are made at extremely low velocities, a velocity regime important to force control. Data collected in a preliminary force control experiment show both a structural flaw in the model and a means to collect data in the extremely slow regime. Figure 6.6 presents the result of an open-loop force move. In this trial, the arm, with a force sensor, was pressed against a hard surface, and the torque is ramped up to a value well above static friction, and down again. The configuration of the arm, actuation and sensor are shown schematically in figure 6.4. A photograph of the apparatus is provided in figure 6.5; by inserting various springs between the arm and cabinet the environmental stiffness could be selected. A number of features are evident in this motion, but one stands out: *there is no stick slip behavior in the resultant force trajectory.* Tustin's model predicts stick slip down to zero velocity. The shaft encoder data from the trial of figure 6.6 are shown in figure 6.7; the motion is extremely slow, about 330 micro-radians (μR) per second. The friction curve in figure 6.6 was calculated according to equation (6.6) below.

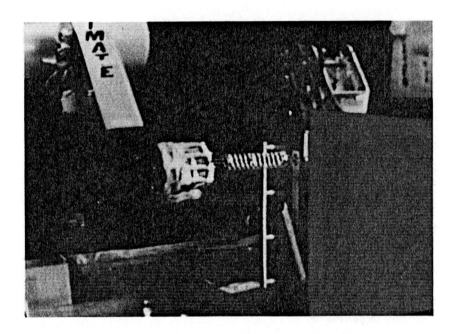

Figure 6.5   PUMA Arm Configured for Compliant Motion Trials.

Figures 6.6 and 6.7 show that by applying varying force between a manipulator and a compliant surface very slow motion can be achieved. The motion velocity is given by:

$$\dot{q} = (1/k) \times \dot{\tau} \qquad\qquad (6.3)$$

where $\dot{q}$ is the motion velocity;
       k is the stiffness of the combined manipulator and environment;
       $\dot{\tau}$ is the torque rate.

Data were acquired using the apparatus of figures 6.4 and 6.5 and a number of coil springs. The springs yielded effective stiffness from 450 to 12,000 N-m per radian. Applying a torque ramp with a rate of 10 N-m per second yielded velocities ranging from 0.0008 to 0.022 radians per second. The study of low speed friction required exceptionally sensitive measurement of velocity and careful accounting of forces. The data analysis led to the observation of the Dahl effect [Dahl 68], which had to be compensated to achieve an accurate measurement of low velocity friction.

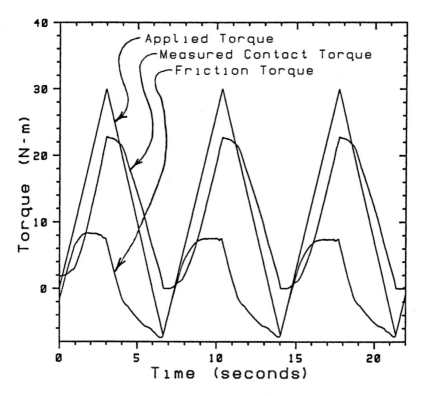

Figure 6.6 Applied Torque and Measured Force (Scaled by Radius) for a Motion in Contact with a Hard Surface (k = 12,400 N-m/rad).

The motion of figures 6.6 and 6.7 shows no stick-slip. Compare these with the motion of figures 6.8 and 6.9, for which a more compliant spring is used and stick-slip results. The limit cycling is evident in both the force and position data. The motion of figures 6.8 and 6.9 is typical of the cases that exhibit stick-slip and are used in the analysis to follow.

Velocity sensing at the level of hundreds of micro-radians per second was necessary and was achieved through careful study of the noise characteristics of the rotational accelerometer and the force sensing fingers. Systron Donner specifies a threshold and resolution for the loaned accelerometer of 0.005% of full scale, substantially below the quantization of the twelve bit A/D converter. The range of the A/D converter and full scale of the instrument are well matched; but when the output signal of the instrument is small, it is possible to make better use of the resolution of the A/D converter by increasing the gain of the analog amplifiers buffering the accelerometer signal. A computer controlled variable gain amplifier permitted an 8x

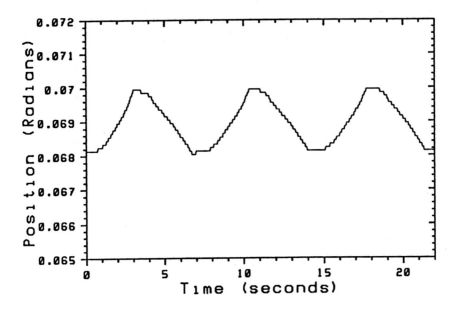

Figure 6.7   Position Reading During the Motion of Figure 6.6.

multiplication of small accelerometer signals thus reducing the effect of quantization. Simpson's rule integration replaced Euler integration in off-line processing to improve the estimate of velocity; and a sensitive bias null was implemented. The combined effect was an order of magnitude reduction in the error in position estimated from the second integral of acceleration, from one part per thousand to one part per ten thousand for large motions.

During compliant motion with constant stiffness, velocity can also be estimated from the derivative of force:

$$\hat{\dot{q}} = (1/k) \times \dot{f} \qquad\qquad (6.4)$$

where $\dot{f}$ is the derivative of measured force.

Note: The friction, applied torque and torsional stiffness are measured in rotational units. Here the Contact forces and stiffness are measured in linear units. Conversion is made by scaling by the appropriate radius, 0.83 meters for most of this work.

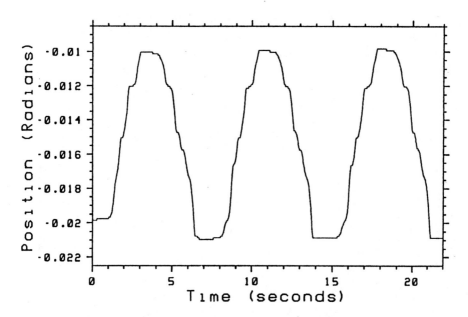

Figure 6.8 Applied Torque and Measured Force (Scaled by Radius) for a Motion in Compliant Contact (k = 2,000 N-m/rad).

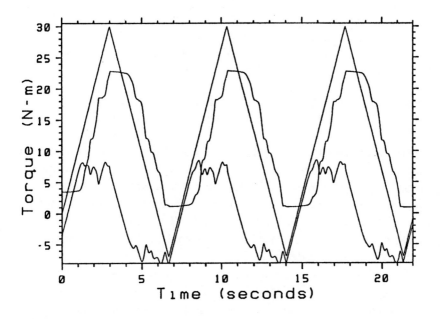

Figure 6.9 Position Reading During the Motion of Figure 6.8.

As the stiffness, k, is large, reasonable force rates correspond to very small velocities. Appropriate preconditioning of the force sensor signal resulted resulted in an rms noise power of 0.01 Newtons, mostly at 60 Hz. The force sensor, however, suffered from considerable slow drift. These effects gave changes in bias in the range of 0.1 Newtons; careful attention to maintaining the null reading of the sensor was required.

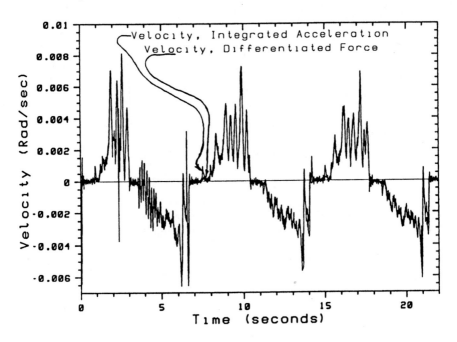

Figure 6.10   Velocity Measured during slow motion by Integration of
                    Acceleration and by Differentiation of Contact Force.

Figure 6.10 shows both the estimate of velocity derived from integrated acceleration and that derived from force rate. The correspondence is remarkable, even at extremely low velocities. The rms difference between the two velocity estimates is 70 μrad/sec, providing assurance that both methods of measurement are quite accurate. The integrated acceleration signal, merged with the differentiated force signal in a cross-over filter, was used as the velocity estimate.

In addition to accurate force and velocity sensing, the measurement of friction at extremely low velocities depended upon accurate accounting of other forces affecting the motion. In the simple model of compliant motions, friction is given by:

$$Friction\ Torque\ =\ Applied\ Torque\ -\ Measured\ Contact\ Torque$$
$$-\ Motion\ Torque. \tag{6.5}$$

The applied torque is known, the contact torque may be determined through scaling the measured contact force by the moment arm; and the motion torque may be computed from the measured accelerations and the known inertia. The resulting estimate of friction force for an early trial is shown in figure 6.11. By combining the velocity and friction data, a plot can be made of friction as a function of velocity, as shown in figure 6.12. At first it was believed that the open curves of figure 6.12 were the result of a hysteresis like process in the friction. Hysteresis is expected due to the frictional lag [Bell and Burdekin 69; Hess and Soom 90]. Simple lack of repeatability was also suspected as the source of the spread. However, study of several different trials frustrated explanation in terms of a sensible, velocity-related process; and examination of several identical trials showed the repeatability to be quite good. Insight came from noting that the spread was greater for motions with softer springs, i.e., those that covered greater distance. Comparison of the spread with the position-dependent friction showed a high correlation. Figure 6.13 is the result of compensating for the position-dependent friction with:

$$Friction\ Torque\ =\ Applied\ Torque\ -\ Measured\ Contact\ Torque$$
$$-\ Motion\ Torque\ -\ Position-Dependent\ Correction. \tag{6.6}$$

Combining the friction measurements of a number of trials with different spring stiffnesses gives figure 6.14, not yet a repeatable measurement of friction as a function of velocity. Again the study of motions carried out with different spring stiffnesses provided the clue to sorting out simultaneous processes. The several curves of figure 6.14 show different angles of rise from the level of friction at zero velocity to a relatively steady friction level at higher velocities. The length of this rise was seen to be variable in terms of time, and variable in terms of velocity, but roughly constant in terms of position change. This observation led to the hypothesis that the Dahl effect was behind the scatter of friction measurements.

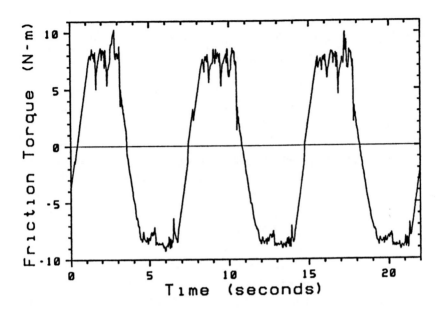

Figure 6.11   Compliant Motion Friction Force Plotted Against Time.

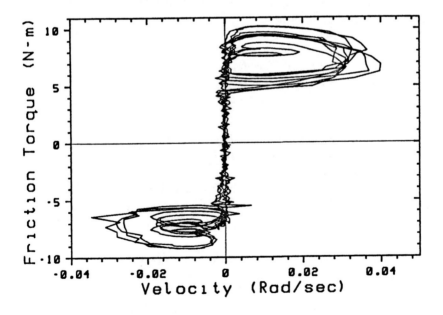

Figure 6.12   Compliant Motion Friction Force Plotted Against Velocity.

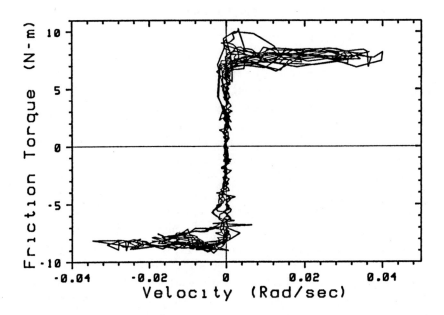

Figure 6.13 Data of Figure 6.12, Corrected According to Equation (6.6).

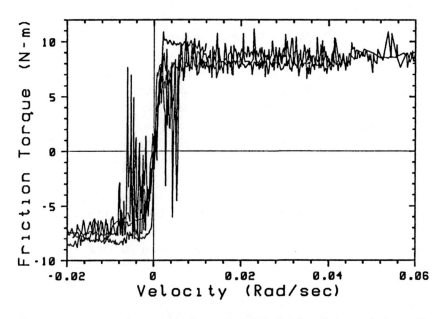

Figure 6.14 Friction Data from Several Trials, Corrected According to Equation (6.6), but not Corrected for the Dahl Effect.

6.4 The Dahl Effect

Dahl's work with friction analysis and modeling is the contemporary work on friction best known to the controls community [Dahl 68, 76, 77]. Studying experimental work with the friction of ball bearings done by H. Shibata at the Aerospace Corporation, Dahl proposed that a process of plastic deformation and failure occurs during the transition from static to kinetic friction, as described in section 2.1. Intuitively the process is this: the asperities of one surface weld to those of the other when the two are in contact for an adequate time. When motion begins the separation is not immediate, but rather the asperities deform in a fashion comparable to that of the bulk material. This process entails both elastic and plastic deformation, and leads to hysteresis if the motion is reversed. Figure 6.15 is taken from [Dahl 77], friction is plotted as a function of displacement - *not velocity* - for a cyclic motion of the bearing. Over the linear region of the curve, that near zero displacement, the process is conservative. A ball bearing, for example, can be observed to rock back and forth on a flat hard surface under the influence of this effect [Dahl 77]. Compare the behavior in the linear region of figure 6.15 with that shown schematically in figure 2.7.

Rabinowicz, [51], reports a brilliantly simple experiment to measure the transition distance from static to kinetic friction. With an apparatus not unlike that of Leonardo da Vinci, Rabinowicz studied the sliding motion of blocks on a flat inclined plane. He would initiate motion by rolling a ball a prescribed distance down the plane to strike the block, thus delivering a known impulse. The relationship between impulse, sliding distance, plate slope and lubricant yield, among other things, information about the transition from static to kinetic friction. For lubricated smooth metal surfaces, the transition distances lie in a range from 1 to 7 microns (μm), roughly the size of the surface asperities (see figure 2.4). Shibata's data show a typical transition distance that is much greater, roughly 200 microns. This inconsistency and Rabinowicz's evidence for transition on a scale too small to be observed in this work, discouraged investigation of the Dahl model. Only when examination of the compliant motion data strongly suggested the presence of the Dahl effect on an observable scale was the possibility studied.

The inconsistency of distance scales is resolved by considering that Rabinowicz was studying the motion of blocks on flat surfaces while Shibata was studying the motion of ball bearings [Rabinowicz 51; Dahl 68]. When a ball is rolling, the relative motion between the ball and contacting surface is much smaller than the motion of the center of the ball. Assuming point contact between the ball and race, a 200 micron motion of a 6 mm ball in a 40 mm race (the dimensions of Shibata's apparatus) corresponds to a

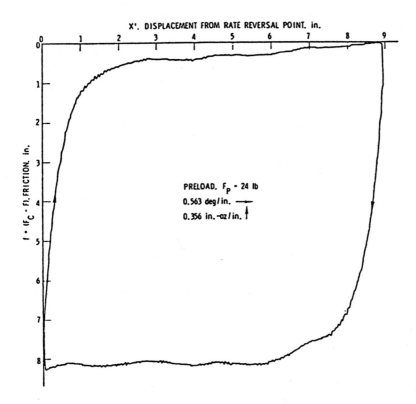

Figure 6.15   Friction Plotted as a Function of Position for a Ball Bearing Apparatus (taken from [Dahl 77], Reprinted Courtesy of Author). Note the Linear Relationship Between Force and Displacement for Small Displacements.

separation of 6.7 microns between opposing faces of the friction interface. If the asperity compliance is comparable in the normal and tangential directions, 6.7 microns is a reasonable figure to compare with Rabinowicz's measurement. The break-away distance for joint 1 of the PUMA arm was computed from the 0.19 meter diameter of the bull gear, assuming a 10% slip coefficient for the involute gear. The externally observable break-away distance was 0.0003 radians. These three measurements of transition distance, shown in table 6.2 all lie near the typical asperity dimension of finished hard metals.

Table 6.2   Break-Away Distance, Transition from Static to Kinetic Friction.

Investigator	Break-Away Distance	Interface Characteristics
E. Rabinowicz [51]	1 to 7 μm	Various Metals Sliding on Steel
P.R. Dahl [77]	6.7 μm	Industrial Ball Bearing, 6mm Balls
B. Armstrong [88]	1.9 μm	Bull Gear of the PUMA Joint 1

6.5 The Stribeck Effect

The Stribeck effect is a name given to decreasing friction with increasing velocity. To observe the Stribeck effect alone, it was necessary to account for the experimental influence of the Dahl effect. Rather than attempting to model the Dahl effect accurately, it was removed from the data by throwing out data taken near to a zero crossing in velocity. Bonding occurs at zero velocity and affects the mechanism behavior for some translation away from the point of bonding. When a sufficient distance is traversed, the Dahl effect is no longer a substantial influence. A number of distances were tried, five shaft encoder ticks (0.0005 radians) was found to be effective. This filtering step, removing all data points collected within five shaft encoder ticks of a velocity zero crossing, I call "de-Dahl'ing".

In the final step of processing, the friction measurements were placed into bins according to velocity. Only those bins containing 30 or more points were retained, a requirement of sufficient sample size to achieve a good estimate of the mean. The result of the "de-Dahl'ing" and sample size filters, applied to the data of figure 6.14, is shown in figure 6.16. The separate lines indicate data collected by different combinations of spring stiffness and torque rate. Binning the data of figure 6.16 according to velocity and computing the mean and deviation give figure 6.17. The negative dependence of friction upon velocity is evident at a level many times the uncertainty in the measurements. The 90% confidence interval shown is taken from the deviation of friction measurements within each bin. The total data set giving figure 6.17 is 4,000 data points, taken in 9 separate motions involving 5 different springs, ranging from 770 to 3400 Newtons per meter stiffness, and 3 different torque rates: 5, 10 and 15 Newton-meters per second. The motion of figures 6.8 and 6.9 is one among the 9 and is typical. Note that in binning friction as a function of velocity, during a limit cycle the velocity will go through a given value twice: once accelerating away from stuck and

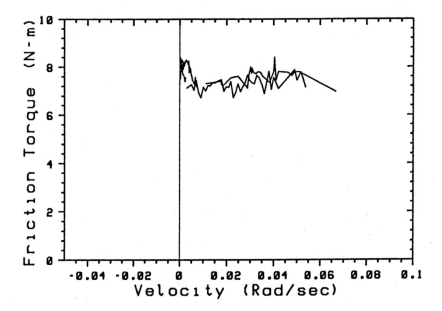

Figure 6.16 Data from Several Trials Corrected for the Dahl Effect.

again accelerating back into stuck. In this analysis these data are combined. The data of figure 6.17 are shown in expanded view in figure 6.18.

Figures 6.17 and 6.18 clearly show a region of negative viscous friction. This is associated with the transition from boundary lubrication to fluid lubrication and is identified as the region of mixed lubrication. Work in tribology on the dynamics of the break-away process and friction in the mixed lubrication regime is active [Zhu and Cheng 88; Sadeghi and Sui 89; Pan and Hamrock 89; Hess and Soom 90]. But for the moment no predictive model of the Stribeck effect is available: an empirical model is required. Several empirical models are examined: an exponential in velocity, $F_s e^{-\dot{x}/\dot{x}_s}$, which is Tustin's model [47]; a Gaussian model, $F_s e^{-(\dot{x}/\dot{x}_s)^2}$; a Gaussian model with offset, $F_s e^{-((\dot{x}-\dot{x}_o)/\dot{x}_s)^2}$; a Lorentzian model, proposed by Hess and Soom [90], $F_s 1/(1 + (\dot{x}/\dot{x}_s)^2)$; and a polynomial model. The models evaluated are presented in table 6.3. The exponential in $\dot{x}$, Gaussian and Lorentzian models, (6.8), (6.10) and (6.13), were also evaluated with two break points giving (6.9), (6.11) and (6.14). I believe that the presence of two break points in the model reflects sliding at two different points in the drive train, for example, at the motor pinion - intermediate gear interface and at the intermediate - bull gear interface. Sliding at different points will give two different effective radii and thus, for the same rate of linear sliding, two different angular rates. The loading and perhaps lubrication at these

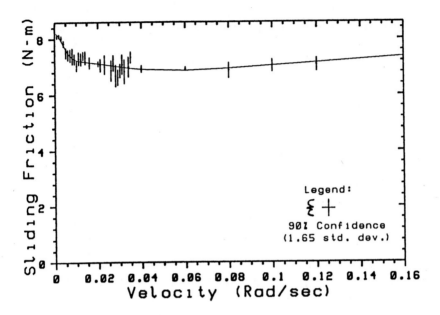

Figure 6.17   Friction as a Function of Velocity During Motions Against a Compliant Surface.

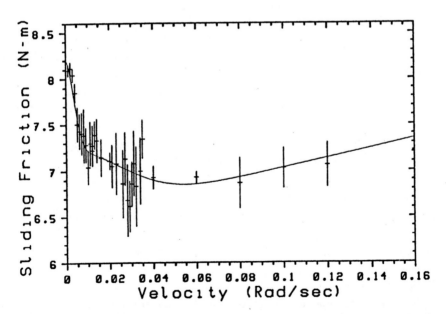

Figure 6.18   Friction as a Function of Velocity During Motions Against a Compliant Surface (Expanded Scale).

Table 6.3 Models Fit to the Friction as a Function of Velocity Data.

Kinetic + Viscous		$F_{cm} + F_{cl} + F_{ol} + F_v \dot{x}$	(6.7)
Exponential in $\dot{x}$	One Break	Kinetic + Viscous + $F_s\, e^{-\dot{x}/\dot{x}_s}$	(6.8)
	Two Breaks	Kinetic + Viscous + $F_{s1}\, e^{-\dot{x}/\dot{x}_{s1}} + F_{s2}\, e^{-\dot{x}/\dot{x}_{s2}}$	(6.9)
Gaussian Model	One Break	Kinetic + Viscous + $F_s\, e^{-(\dot{x}/\dot{x}_c)^2}$	(6.10)
	Two Breaks	Kinetic + Viscous + $F_{s1}\, e^{-(\dot{x}/\dot{x}_{s1})^2} + F_{s2}\, e^{-(\dot{x}/\dot{x}_{s2})^2}$	(6.11)
With Offset	Two Breaks	Kinetic + Viscous + $F_{s1}\, e^{-((\dot{x}-\dot{x}_{o1})/\dot{x}_{s1})^2} + F_{s2}\, e^{-((\dot{x}-\dot{x}_{o2})/\dot{x}_{s2})^2}$	(6.12)
Lorentzian Model	One Break	Kinetic + Viscous + $F_s\, \frac{1}{1+(\dot{x}/\dot{x}_s)^2}$	(6.13)
	Two Breaks	Kinetic + Viscous + $F_{s1}\, \frac{1}{1+(\dot{x}/\dot{x}_{s1})^2} + F_{s2}\, \frac{1}{1+(\dot{x}/\dot{x}_{s2})^2}$	(6.14)
Polynomial in $\dot{x}$		Kinetic + Viscous + $F_2\, \dot{x}^2 + F_3\, \dot{x}^3 + F_4\, \dot{x}^4 + F_5\, \dot{x}^5 + F_6\, \dot{x}^6 + F_7\, \dot{x}^7$	(6.15)

points may also be different, giving rise to different characteristic velocities of the Stribeck curve [Hess and Soom 90].

The models were fit to a data set that merged data from the compliant motions, the closed-loop, constant-velocity motions and the open-loop, constant-torque motions. The values of kinetic friction demonstrated by these experiments were observed to be different, and three different kinetic friction parameters were introduced into the model as shown in equation (6.7). Because they were identified with each model fitting process, these three kinetic friction parameters could potentially vary with the proposed model. But in fact they were observed to be consistent to within a percent across the eight models examined. For the two break Lorentzian model, (6.14), the three kinetic friction parameters were:

Compliant Motion:	$F_{cm} = 6.40$ [N-m];
Closed-Loop Constant-Velocity Motion:	$F_{cl} = 7.81$ [N-m];
Open-Loop Constant-Torque Motion:	$F_{ol} = 8.42$ [N-m].

The source of this observed difference in the kinetic friction parameter is unknown. The data were collected at different times; drift in the kinetic friction parameter was observed and is discussed in relation to adaptive control, but it was generally less than 10% or 0.8 N-m. The data were also collected by different experiments; an experimental bias affecting the compliant motion data is indicated by the large shift in kinetic friction relative to the constant-torque and constant-velocity motion data. The difference in kinetic friction is particularly surprising between the closed-loop, constant-velocity motions and the open-loop, constant-torque motions. These data are average torque at constant velocity and average velocity at constant torque - data that do not require the extensive processing of the compliant motion data. The merged friction data set comprises 38 data points and is shown in figure 6.19, which spans the velocity range of joint one.

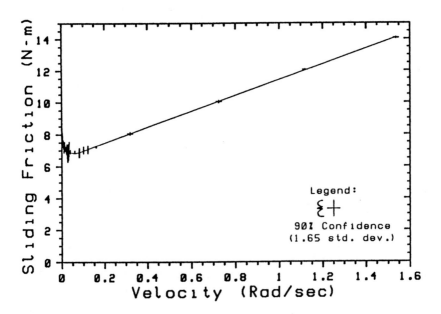

Figure 6.19   Friction as a Function of Velocity; Compliant, Constant Velocity and Constant Torque Motion data Merged to Span Full Motion Range.

Linear parameters of the several models were fit to the data using the variance weighted normal equations, (3.2). Non-linear parameters were fit by exhaustive search, which is to say that the goodness of fit was evaluated at every point on a mesh and the best fit selected. Exhaustive search was practical because the dimensionality of the problem is low and because the computation need only be performed once. Local minima were observed, posing a challenge to gradient techniques. Analysis of variance data for the nine models are presented in table 6.4.

Table 6.4 Analysis of Variance Data for Several Empirical Models of Stribeck Friction. DOF: Degrees of Freedom

Model Name		DOF of Model	Residual Variance, Weighted	DOF of Residual	MSE per DOF
Kinetic + Viscous		4	819.26	34	24.10
Exponential in $\dot{x}$	One Break	6	94.48	32	2.95
$[e^{-(\dot{x}/\dot{x}_\bullet)}]$	Two Breaks	8	63.88	30	2.13
Gaussian Model	One Break	6	154.42	32	4.83
$[e^{-(\dot{x}/\dot{x}_\bullet)^2}]$	Two Breaks	8	47.21	30	1.57
With Offset	Two Breaks	10	35.26	28	1.26
Lorentzian Model	One Break	6	93.04	32	2.91
$[\frac{1}{1+(\dot{x}/\dot{x}_\bullet)^2}]$	Two Breaks	8	48.88	30	1.63
Polynomial in $\dot{x}$		10	92.01	28	3.29

The weighted residual variance and mean squared error per degree of freedom are the significant figures of table 6.4. These indicate the goodness of the model fit. The units of the weighted residual variance are sum squared standard deviations, so computing the MSE gives the mean number of standard deviations by which the model misses each datum. If the errors were independent and the model were exact, one would expect an MSE of one, or one standard deviation of difference between the model and data per residual degree of freedom. There are several fits that approach this value. For comparison, the error bars in figures 6.17 and 6.19 indicate the 90% confidence intervals, or 1.65 standard deviations. The fit of the kinetic plus viscous friction model is provided as a basis for estimating the improvement.

The data of Table 6.4 show that the models to consider are the two break Gaussian model, (6.11), Lorentzian model, (6.14), and Gaussian model with offset, (6.12). It is not surprising that the Gaussian and Lorentzian models do comparably well: expanded in power series, $e^{-(\dot{x}/\dot{x}_*)^2}$ and $\frac{1}{1+(\dot{x}/\dot{x}_*)^2}$ are identical to the quartic term. When plotted these models differ only in their tails, where the Gaussian model dies out more rapidly. The difference in weighted residual variance, 47.21 vice 48.88, is certainly not enough to justify the choice of one model over the other. The choice should be made on the pragmatic basis of suitability to purpose. As neither gives a dynamic for which the differential equation of system motion can be solved, even here we are left with little distinction.

Using the two break Gaussian model and minimizing the weighted squared error yields:

$$F(\dot{x}) = 8.45 + 4.92\,\dot{x} + 1.06\,e^{-(\dot{x}/0.0061)^2} + 0.59\,e^{-(\dot{x}/0.048)^2} \qquad (6.16)$$

where the kinetic friction parameter of the open-loop motion is presented. Equation (6.16) is plotted as the model curve in figures 6.17, 6.18 and 6.19.

Examination of figure 6.17 suggests that there should be an offset in the location of the lower break point, as given by equation (6.12). A shift to the right of the break point, coupled with increasing the steepness of the break, might yield a better fit. The Gaussian model with offset,

$$F(\dot{x}) = 8.45 + 4.92\,\dot{x} + 0.81\,\xi(-(\frac{\dot{x}-0.0022}{0.0030})^2) + 1.98\,\xi(-(\frac{\dot{x}+0.100}{0.104})^2),$$
$$\qquad (6.17)$$

$$\text{where} \quad \xi(x) \;=\; \begin{cases} 1, & \dot{x}-\dot{x}_o \le 0; \\ e^x, & \dot{x}-\dot{x}_o > 0. \end{cases}$$

yielded the weighted residual variance of 35.26, which is significantly lower the 47.21 of the Gaussian model. Adding the two offset parameters gives a Fisher statistic of $F(2,28) = 4.74$. Assuming randomness in the errors, $F(2,28)$ greater than or equal to 4.74 supports accepting the affirmative hypothesis at the 97% confidence level. Thus, use of this model is justified on purely statistical grounds. But physically it is quite unsatisfactory in that one break receives a shift to the right, $(\dot{x} - 0.0022)$, and the other to the left, $(\dot{x} + 0.100)$. If the addition of the offset in fact describes a physical process, we would expect the two shifts to appear in the same direction. For this reason the empirical model is rejected.

The final model of tables 6.3 and 6.4 is a 7th order polynomial in $\dot{x}$. The matrix inversion of the normal equation, (3.2), became singular to the precision of 64 bit floating point for polynomials greater than 7th order. Gradient search or other algorithms could be used to fit larger polynomials, but the effort is not indicated. By 7th order, the polynomial model forms two minima and seems to be contorting itself to fit what appear to be outliers in figure 6.18.

In the analysis of chapter 7, the Lorentzian model will be employed. The added complexity of the offset Gaussian model is not physically justifiable, and the Lorentzian model will more conveniently support the analysis. The model, with identified parameters is:

$$F(\dot{x}) = 8.42 + 4.94\,\dot{x} + 1.30\,\frac{1}{1 + (\dot{x}/0.0058)^2} + 0.466\,\frac{1}{1 + (\dot{x}/0.068)^2} \quad (6.18)$$

The Stribeck effect has profound implications for the stability of controlled mechanisms. In the region of negative viscous friction the system is quite unstable. The derivative of the friction model of equation (6.16) is shown in figure 6.20. The derivative of friction with respect to velocity is the natural damping of the system: a negative derivative indicates an unstable system. For the two control structures tested - stiff velocity feedback and direct accelerometer feedback - the lowest velocities achievable before the onset of stick-slip are indicated in figure 6.20. The degree of instability that must be compensated at the lowest point of figure 6.20 is more than four times as great as that achieved by direct accelerometer feedback. Among grease and oil lubricated mechanisms and current control instrumentality and technique, the Stribeck effect gives rise to a *minimum* velocity for stable motion and a *minimum* step that may be taken. The nature of these minima

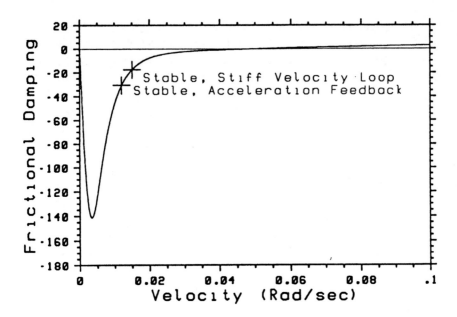

Figure 6.20   The Derivative of the Friction Model of Equation (6.16), in [N-m/rad/sec].

and the control performance necessary to compensate Stribeck friction are
the subject of the next chapter.

6.6 Temporal Effects in the Rise and Decay of Friction

Friction is not a simple function of velocity. Two temporal phenomena
in changing friction are introduced in chapter 2. The first is a time lag in
the change in friction corresponding to a change in velocity, called frictional
lag [Hess and Soom 90; Bell and Burdekin 66, 69; Rabinowicz 58; Rice
and Ruina 83]. Frictional lag is illustrated schematically in figure 2.19 and
empirically in figures 2.20, 2.21 and 2.22. The second temporal phenomenon
is a dependence of static friction on dwell time at zero velocity, [Bell and
Burdekin 66; Brockley and Davis 68; Rabinowicz 65; Kato *et. al.* 72]. This
effect is illustrated in figure 2.17.

These phenomena are evident in the friction data of the PUMA
manipulator, but it is necessary to average considerable lengths of data to
reduce the impact of position-dependent friction and mechanism flexibility.
That is to say that the phenomena are not clearly evident in the *instantaneous*
friction-velocity data which were used to develop the Stribeck curve. The
phenomena are evident in the instantaneous data of Hess and Soom, [90],

and Bell and Burdekin, [69], who employ stiff apparatus and controlled sliding conditions, as opposed to a practical servo mechanism.

By averaging lengths of data and observing the relation between modeled friction and measured friction the frictional lag can be measured. Figure 6.21 is a plot of the correlation between the friction model of equation (6.18), and the experimental data of one of the compliant motion trials. The model is evaluated with a frictional lag ranging from zero to 60 milli-seconds (mS), as described by equation (2.3). The peak correlation in figure 6.21 occurs at 48 mS of delay. The correlations were formed by applying the de-Dahl'ing filter to the data and combining only data points for which both the velocity and friction were measured during motion. Combining data from 9 trials gives:

$$\tau_L = 0.046 \quad [\text{sec}] \qquad\qquad 6.19$$

with a standard deviation of $\sigma_{\tau_L} = 0.013$ [sec]. The standard deviation is quite large, and the variations are uncorrelated with stiffness or velocity. There are two samples which show a negative correlation time and are probably outliers. When these are rejected the mean rises to $\tau_l = 0.051$ and the standard deviation is reduced by half.

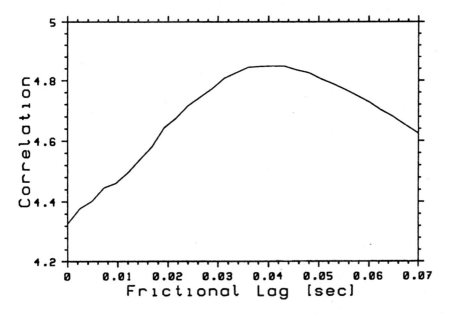

Figure 6.21   Correlation Between the Friction Model of Equation (6.18) and the Friction Data with Frictional Lag Introduced.

This value of frictional lag is substantially greater than those observed by Hess and Soom, [90], which fall in the range of 3 - 10 mS; or the figure of 3 mS observed by Xiaolan and Haiqing [87]. Hess and Soom determine that frictional lag rises with increasing contact loading and lubricant viscosity. As gear teeth are among the most highly loaded sliding contacts and the lubricant was a PD-0 grease,† a longer lag may be expected.

Figure 6.22 is a plot of the difference between the maximum and minimum friction during a stick-slip cycle, plotted as a function of the dwell time. Dwell time and the stick-slip amplitude are defined in figure 2.15. Figure 2.17 shows the static friction to rise as a function of time. Note the logarithmic time scale. There is evidence, [Rabinowicz 58], that the rise of static friction over kinetic friction goes to zero as the dwell time goes to zero; but dwell times shorter than one tenth of a second could not be observed here.

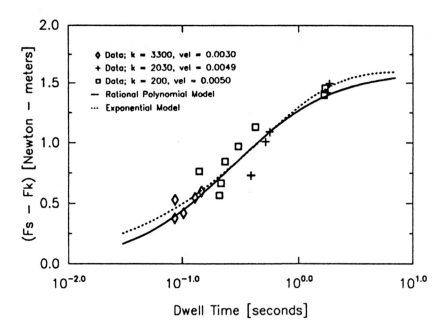

Figure 6.22 $(F_s - F_k)$ as a Function of Dwell Time, t_2. Solid curve given by $1.61(\frac{t_2}{t_2 + 0.264})$ [Derjaguin *et. al.* 57]; dashed curve given by $1.61(1 - e^{1.66 t_2^{0.65}})$ [Kato *et. al.* 72].

---

† Optimal PD-0 is specified for the PUMA arm. The PD numbers are part of the AGMA grease specification system, PD-0 indicating semi-fluid consistency.

The solid curve of figure 6.22 is a fit to the model of Derjaguin *et. al.* , [57]:

$$(F_s - F_k) = \lambda \frac{t_2}{t_2 + \gamma} \tag{6.20}$$

where λ is the magnitude of the difference between static and kinetic friction, in this case 1.61 [N-m], and γ is the characteristic time of the static friction rise. Kato *et. al.* [72] employ and justify a different model, presented in equation (2.2) and shown as the dashed curve in figure 6.22. The two models give curves of comparable form, but the model of Derjaguin *et. al.* is mathematically more suitable to the purposes of the next chapter and for this reason is employed.

6.7 Variance in Friction as Process Noise

Estimation of the repeatability of the friction forces is a basic object of this report. During motion one would expect the friction force to be a combination of systematic and stochastic contributions. If the stochastic contribution has a correlation time that is short relative to the time constants of the dynamic system, it may be modeled as a white noise disturbance. From this we would expect that the power density spectrum of the sliding process noise to be flat, giving a constant noise power per Hertz. This process noise will contribute to variance in the system state and will reduce the performance of state estimators, increasing the difficulty of stabilizing such a system by feedback control.

Figures 6.23, 6.24 and 6.25 are power spectra of the acceleration signal at average velocities of zero, 0.00081 and 0.795 radians per second respectively. The figures show an increase in the broad band power in the acceleration signal with increasing velocity. The signal of the first figure is the response of the arm to ambient vibrational noise. It is included to provide a baseline against which the other two signals may be compared. 10^{-9} [rad/sec]^2/Hertz is the level of quantization noise in the measurement process. The large peak at 18 Hertz corresponds to the first resonant mode of the arm. The smaller peak, a factor of 20 lower in power, may correspond to a resonant mode of the arm plus supporting base.

Figures 6.24 and 6.25 present the same curve computed from motion data. The motion of figure 6.24 was a compliant motion with a stiffness of 12,343 [N-m/rad], the motion of figures 6.6 and 6.7. The motion of figure 6.25 was an unconstrained, open-loop, constant torque motion, like that presented in figure 4.1.

With the assumption that noise in the acceleration signal is dominated by frictional disturbance, the friction noise power may be estimated from the

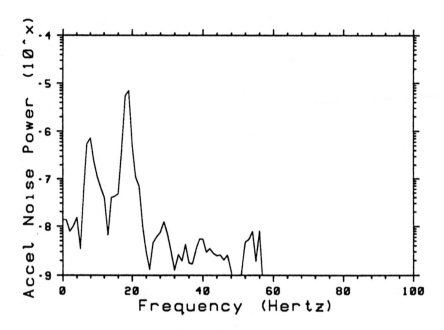

6.23   Power Spectrum of the Accelerometer Signal, $\dot{\theta}_{\text{ave}} = 0$.

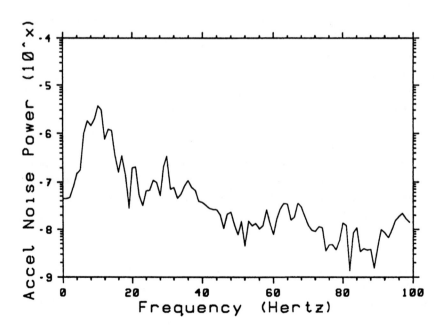

6.24   Power Spectrum of the Accelerometer Signal, $\dot{\theta}_{\text{ave}} = 0.00081$.

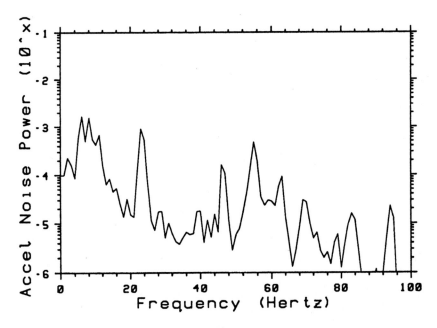

6.25   Power Spectrum of the Accelerometer Signal, $\dot{\theta}_{ave} = 0.795$. (Note the adjusted scale of the ordinate relative to figures 6.23 and 6.24.)

measured acceleration noise. To avoid contribution by the flexible response of the system, the friction noise power is estimated by averaging the power per Hertz across the range from 20 to 50 Hertz. The accelerometer has a second order roll off at 40 Hertz, for which a correction was applied.

Forming the average and scaling by the 6.3 [kg-m²] inertia of the arm gives $\sigma_F^2 = 4.2x10^{-6}$ [(N-m)²/Hz] at 0.00081 [rad/sec] and $\sigma_F^2 = 3.9x10^{-3}$ [(N-m)²/Hz] at 0.795 [rad/sec]. Compared with friction values of 10 [N-m], the friction noise power is not very great.

In this chapter the relationship between friction, velocity and system state has been explored. In the moderate to high velocity regime, friction is seen to be linear with velocity and described by separate kinetic and viscous friction parameters in the positive and negative motion directions. Low velocity friction is explored and the dependence of friction upon displacement, velocity, lag time and dwell time is mapped out. The interaction of these frictional phenomena and feedback control will be studied in the next chapter.

Chapter 7

Analysis of Stick-Slip

When moving slowly machines are likely to exhibit stick-slip, a periodic cycle of alternating motion and arrest. Stick-slip determines the lower performance bounds of a machine, the lowest sustainable speed and the shortest governable motion. In applications that place a premium on precise motion, an understanding of the dynamics of stick-slip and its possible elimination take on great value.

Stick-slip is governed by four phenomena:

- The mass-spring (system-control) dynamics of the system;
- The non-linear, low velocity friction;
- The interdependence of the static friction and dwell time;
- The time lag between a change in system state and the corresponding change in friction.

The system to be analyzed consists of a sliding mass governed by a proportional-derivative (PD) controller. The sliding contact friction combines Stribeck friction, frictional lag and dwell dependent static friction. The system is similar to that shown schematically in figure 2.14. In analysis, the dynamics of ideal PD control are interchangeable with the those of a mass-spring system with damping. The Stribeck friction is described by the Lorentzian model, equation (6.13); the frictional lag by equation (2.3), and the dwell dependent static friction by equation (6.20). The combined dynamic is expressed:

$$M\ddot{x} = -k_p\,x - k_v\,\dot{x} - F_k\,\mathrm{sgn}\,(\dot{x}) - F_v\,\dot{x} - \lambda\,(\gamma,t_2)\,f_s\,(\dot{x}/\dot{x}_s,\tau_L)\,\mathrm{sgn}\,(\dot{x}) \quad ; \quad (7.1)$$

$$f_s\,(\dot{x}/\dot{x}_s,\tau_L) = \begin{cases} \dfrac{1}{1 + (\dot{x}\,(t - \tau_L)\,/\dot{x}_s)^2}, & t > \tau_L \\ 1, & t \le \tau_L \end{cases} \qquad (7.2)$$

The symbols used in this chapter are defined below. The friction model is made up of F_k and F_v, the customary kinetic and viscous friction terms, and $\lambda(\gamma,t_2)f_s(\cdot)$, which models the Stribeck friction. The Stribeck friction model is divided into two components: the magnitude, $\lambda(\gamma,t_2)$, which will depend on the dwell time and has dimensions of force, and a velocity dependent term, $f_s(\cdot)$, which is dimensionless, ranges from zero to one, and exhibits the delay given by τ_L. Sliding is taken to begin at $t = 0$. The case statement in the definition of $f_s(\cdot)$ is used to reflect the fact that $f_s(\cdot) = 1$ prior to sliding. The physical dimensions of linear sliding are used here: with appropriate changes of dimension rotational motion may be considered by the same analysis.

—————————————— *Nomenclature* ——————————————

[units given for translational motions]

$$M = \text{mass, [kg]}$$
$$k_p = \text{stiffness, [Newtons/m]}$$
$$k_v = \text{derivative feedback or damping, [Newtons/m s}^{-1}]$$
$$k_i = \text{integral control term, [Newtons/m s]}$$
$$k_i^* = \text{dimensionless integral control term; } k_i^* = (k_i/M)/\beta^3$$
$$F = \text{total friction force, [Newtons]}$$
$$F_k = \text{kinetic friction, [Newtons]}$$
$$F_k^* = \text{dimensionless kinetic friction; } F_k^* = \frac{F_k/M}{\alpha\beta^2}$$
$$F_v = \text{viscous friction, [Newtons/m s}^{-1}]$$
$$k_v' = k_v + F_v; \text{ merged damping and viscous friction}$$
$$\omega = \sqrt{k_p/M}; \text{ the natural frequency, [s}^{-1}]$$
$$\rho = k_v'/2\omega; \text{ the damping factor, [·]}$$
$$x,\dot{x},\ddot{x} = \text{position, velocity and acceleration in physical coordinates, [m], etc.}$$
$$y,\dot{y},\ddot{y} = \text{position, velocity and acceleration in physical coordinates,}$$
$$\text{shifted position, [m], etc.}$$
$$\xi,\dot{\xi},\ddot{\xi} = \text{position, velocity and acceleration in dimensionless coordinates, [·]}$$
$$\xi_b = \text{dimensionless position at the moment of break-away}$$
$$\xi_a = \text{dimensionless position at the moment of arrival in static friction}$$
$$\dot{x}_d = \text{desired sliding velocity, [m s}^{-1}]$$
$$\dot{\xi}_d = \text{dimensionless desired velocity; } \dot{\xi}_d = \dot{x}_d/\dot{x}_s, [\cdot]$$
$$\dot{x}_s = \text{characteristic velocity of the Stribeck friction, [m s}^{-1}]$$
$$\alpha = \dot{x}_s/\omega; \text{ scaling factor for length, [m]}$$
$$\beta = \omega, \text{ scaling factor for time, [s}^{-1}]$$
$$t = \text{time, [s]}$$
$$t_1 = \text{duration of the slip portion of the stick-slip cycle, [s]}$$
$$t_2 = \text{duration of the stick portion of the stick-slip cycle, dwell time, [s]}$$
$$\gamma = \text{characteristic time of the static friction rise, [s]}$$
$$\tau_L = \text{magnitude of the time lag in sliding friction, [s]}$$
$$t^*,t_2^* = t, t_2, \text{ in dimensionless coordinates, [·]}$$

Equation (7.1) is a continuous, non-linear differential equation, whose behavior is complicated by the fact that $\lambda(\gamma, t_2)$ is not constant, but is itself a function of γ and the dwell time, t_2. The term $\lambda(\gamma, t_2)$ is used to model the rise of static friction with dwell time, a process outlined in section 2.3. Motivation for the model governing $\lambda(\gamma, t_2)$ is described in section 7.3, where more of the pieces are available to describe the interaction of $\lambda(\gamma, t_2)$ and the stick-slip cycle. Deferring the details, $\lambda(\gamma, t_2)$ is governed by a sampled process in which the magnitude of static friction at the beginning of each stick-slip cycle is given by:

—————————————— *Nomenclature* ——————————————

$\gamma^*, \tau_L^* =$		γ, τ_L, in dimensionless coordinates, [·]
$\vec{r} =$		radius vector of a trajectory in the phase plane, figure 7.2
$E_k^* =$		dimensionless kinetic energy, [·]
$\Delta_{E_k^*} =$		change in dimensionless kinetic energy during slip, [·]
$\Delta_{E_p^*} =$		energy disipated during a slip cycle by damping, [·]
$\Delta_{E_w^*} =$		energy contributed to a slip cycle by spring windup, due to static friction, [·]
$\Delta_{E_{f_\bullet}^*} =$		perturbation to the slip energy due to Stribeck friction, [·]
$\Delta'_{E_{f_\bullet}^*} =$		$\Delta_{E_{f_\bullet}^*}$ scaled by $\lambda_b^* \dot{\xi}_d$, [·]
$t_b =$		time at the moment of break-away, [s]
$t_a =$		time at the moment of arrival in static friction, [s]
$t_b^*, t_a^* =$		t_b, t_a in dimensionless coordinates, [·]
$\lambda(\gamma) =$		instantaneous magnitude of the static and Stribeck friction, [Newtons]
$\lambda_\infty =$		steady state magnitude of the static and Stribeck friction, [Newtons]
$\lambda_b =$		magnitude of the static friction at t_b, [Newtons]
$\lambda_a =$		magnitude of the static friction at t_a, [Newtons]
$\lambda_{b_n} =$		λ_b on the n_{th} stick-slip cycle, [Newtons]
$\lambda_{a_{n-1}} =$		λ_a on the $(n-1)^{th}$ stick-slip cycle, [Newtons]
$\lambda_{b_{\bullet\bullet}} =$		steady state magnitude of the static and Stribeck friction, [Newtons]
$\lambda^*, \lambda_\infty^*,$		
$\lambda_b^*, \lambda_a^* =$		corresponding friction magnitudes scaled to
$\lambda_{b_n}^*, \lambda_{a_{n-1}}^*$		dimensionless coordinates, [·]
$h =$		static friction reduction factor, [·]
$C_F =$		constant of integration for integral control, [·]
$C_{\bar{0}} =$		constant of integration for integral control, [·]
$F_{ib}^* =$		dimensionless force applied by integral control at break-away, [·]
$\text{sgn}(\cdot) =$		the signum function

——————————————

Subscripts:	b	marks a variable given at the moment of break-away
	a	marks a variable given at the moment of arrival in static friction
Superscripts:	$*$	indicates a dimensionless quantity

$$\lambda_{b_n} = \lambda_{a_{n-1}} + \left(\lambda_\infty - \lambda_{a_{n-1}}\right) \frac{t_2}{t_2 + \gamma} \qquad (7.3)$$

where λ_{b_n} is the magnitude of the static friction during the nth slip cycle; $\lambda_{a_{n-1}}$ is the level of Stribeck friction at the moment of entering the stuck condition on the previous cycle; t_2 is the time spent in static friction, that is the dwell time; λ_∞ is the ultimate level of static friction, attained as $t_2 \to \infty$, and γ is the time scale of the rise of static friction to λ_∞.

Thus the system under study is modeled as a non-linear differential equation in the system state, (7.1), coupled to a non-linear difference equation in the magnitude of static friction, (7.3). The combined system is one of 9 parameters, M, k_p, k_v, λ_∞, γ, $\dot{x}_s$, τ_L, F_k, and F_v. We shall consider the situation in which the desired motion is steady, i.e., $\dot{x}_d$ is a constant, creating a tenth parameter. The object of this chapter is to determine the range of these parameters over which such a system will exhibit stick-slip.

7.1 Dimensional Analysis

We investigate steady motion, i.e. $\ddot{x}_d = 0$, of a system with kinetic plus viscous plus Stribeck friction and with proportional plus derivative feedback, as described by equations (7.1), (7.2) and (7.3). Dimensional analysis will provide a reduction in the number of independent parameters from 10 to 5, yielding a simpler model and greatly facilitating the perturbation analysis to follow.

The first transformation is a shift of the position coordinate to place the equilibrium point at the origin. Physically, this shift corresponds to finding the point at which the proportional term balances kinetic friction and the damping force due to steady motion. The shift eliminates the parameter F_k from the dynamic equation. Folding the viscous friction parameter into the derivative feedback, one may write $k_v' = k_v + F_v$.

Defining $\qquad y = x - x_d + \{F_k + k_v'\dot{x}_d\}/k_p$;

then $\qquad\qquad \dot{y} = \dot{x} - \dot{x}_d$, $\qquad\qquad\qquad\qquad\qquad\qquad (7.4)$

$\qquad\qquad\quad \ddot{y} = \ddot{x}$.

(Note: $\ddot{y} = \ddot{x}$ because $\ddot{x}_d = 0$, only steady desired motion is considered.)

With this shift of the position coordinate, equation (7.1) becomes:

$$M\ddot{y} = -k_p y - k_v'\dot{y} - \lambda\left(\gamma, t_2\right) f_s\left((\dot{y} + \dot{x}_d)/\dot{x}_s, \tau_L\right) . \qquad (7.5)$$

The sgn($\dot{x}$) functions are eliminated: when the commanded motion, $\dot{x}_d$, is steady motion in one direction, the velocity will not reverse under quite general circumstances [Derjaguin *et. al.* 57]. Equation (7.5), coupled with equation (7.3), has 8 independent parameters, $\dot{x}_d$ included, in 3 physical dimensions: mass, length and time. The Buckingham Pi theorem indicates that the system may be described by (8 - 3) dimensionless groups. Scaling the position coordinate by a factor α and time coordinate by a factor $1/\beta$ gives dimensionless position and time, ξ and t*:

$$\alpha\xi \equiv y \qquad \text{and} \qquad t^* \equiv \beta t . \qquad (7.6)$$

Forming the derivatives of ξ w.r.t. the new time coordinate, we find that

$$\frac{dy}{dt} = \alpha\beta\frac{d\xi}{dt^*} \qquad \text{and} \qquad \frac{d^2y}{d^2t} = \alpha\beta^2\frac{d^2\xi}{d^2t^*} . \qquad (7.7)$$

Substituting ξ for y and its derivatives in (7.5) gives:

$$M\alpha\beta^2\ddot{\xi} = -k_p\alpha\xi - k_v'\alpha\beta\dot{\xi} - \lambda(\gamma^*,t_2^*)\, f_s\left(\alpha\beta\left(\dot{\xi}+\dot{\xi}_d\right)/\dot{x}_s, \tau_L^*\right) , \qquad (7.8)$$

where $\dot{\xi}_d$ is the magnitude of the desired velocity in dim'less coordinates;
$\alpha\beta\dot{\xi}_d = \dot{x}_d$, $\gamma^* = \beta\gamma$, and $\tau_L^* = \beta\tau_L$;
(Note: $\dot{\xi} = 0$ is the desired dimensionless velocity, not $\dot{\xi} = \dot{\xi}_d$).

Dividing through by the mass yields a system with scaled force constants. Writing (7.8) in terms of ω and ρ, the customary second order system parameters, and dividing through by $\alpha\beta^2$ gives:

$$\ddot{\xi} = -\frac{\omega^2}{\beta^2}\xi - 2\rho\frac{\omega}{\beta}\dot{\xi} - \frac{\lambda(\gamma^*,t_2^*)/M}{\alpha\beta^2}f_s\left(\alpha\beta\left(\dot{\xi}+\dot{\xi}_d\right)/\dot{x}_s, \tau_L^*\right) \qquad (7.9)$$

By choosing the time scale, β, to eliminate ω and the length scale, α, to eliminate $\dot{x}_s$, the dimensionless dynamic model is formed.

$$\beta = \omega = \sqrt{k_p/M} \qquad \text{and} \qquad \alpha = \frac{\dot{x}_s}{\beta} = \frac{\dot{x}_s}{\omega} , \qquad (7.10)$$

$$\ddot{\xi} = -\xi - 2\rho\dot{\xi} - \lambda^*\left(\gamma^*, t_2^*\right) f_s\left(\left(\dot{\xi} + \dot{\xi}_d\right), \tau_L^*\right) \quad , \tag{7.11}$$

where $\rho = (k_v'/M)/2\omega$.

$$f_s\left(\left(\dot{\xi} + \dot{\xi}_d\right), \tau_L^*\right) = \begin{cases} \frac{1}{1 + (\dot{\xi}(t^* - \tau_L^*) + \dot{\xi}_d)^2}, & t^* > \tau_L^* \\ 1, & t^* \leq \tau_L^* \end{cases} \tag{7.12}$$

$$\lambda_{b_n}^* = \lambda_{a_{n-1}}^* + \left(\lambda_\infty^* - \lambda_{a_{n-1}}^*\right) \frac{t_2^*}{t_2^* + \gamma^*} \tag{7.13}$$

where $\lambda^*\left(\gamma^*, t_2^*\right) = \frac{\lambda(\gamma^*, t_2^*)/M}{\alpha\beta^2} = \frac{\lambda(\gamma^*, t_2^*)/M}{\omega\dot{x}_s}$.

The dimensionless model is given by equations (7.10) - (7.13), a nonlinear mixed system in the five parameters ρ, λ_∞^*, $\dot{\xi}_d$, τ_L^*, and γ^*. By setting $\beta = \omega$ time measurements are scaled to units of the natural frequency of the system. Thus as stiffness increases, perhaps due to an increase in control gain, τ_L^* and γ^* increase as well. By scaling lengths according to $\alpha = \dot{x}_s/\omega$, the characteristic velocity of the Stribeck friction becomes the unit velocity. Thus systems are scaled so that the range of motion velocity influenced by Stribeck friction, a range which will vary between systems, will correspond to dimensionless velocities of a fixed range Roughly speaking, the dimensionless velocities influenced by Stribeck friction lie roughly the range from zero to ten. The dimensionless system has unit stiffness, that is k_p^*, if it were considered, would always equal one. As physical stiffness changes both α and β change, and thus the scaling is changed for all of the parameters and state variables.

In the dimensionless coordinates energy is proportional to radius on the phase plane, $(\xi^2 + \dot{\xi}^2)$ (when discussing the energy of a system under feedback control, the proportional term is treated as a virtual spring).

Writing in physical coordinates, the system energy is given by:

$$E = \frac{1}{2}k_p x^2 + \frac{1}{2}M\dot{x}^2 = \frac{1}{2}k_p \frac{\dot{x}_s^2}{\omega^2}\xi^2 + \frac{1}{2}M\dot{x}_s^2\dot{\xi}^2 = \frac{1}{2}M\dot{x}_s^2\left(\xi^2 + \dot{\xi}^2\right) \quad . \tag{7.14}$$

Energy in dimensionless coordinates is given by:

$$E^* = \frac{1}{2}\left(\xi^2 + \dot{\xi}^2\right) \quad . \tag{7.15}$$

The derivation of the dimensionless model does not depend on the structure of either non-linear term, except to the extent that the Stribeck friction is modeled as a function of $(\dot{x}/\dot{x}_s)$. Thus, as tribological understanding improves and new friction models are presented, this dimensional analysis can be adapted.

7.2 Perturbation Analysis

Equations (7.10) - (7.13) present the coupled differential and difference equations governing the low velocity motion of a servo controlled mechanism. No analytic solution of equation (7.12) is available. And while simulation is a useful tool in specific instances, the model has five independent parameters, creating a space of systems too large to fully explore numerically. To determine the parameter combinations that signal the onset of stick-slip motion, a perturbation technique will be used.

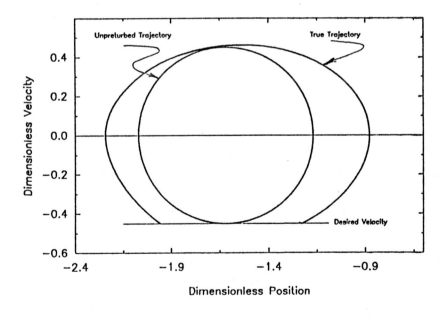

Figure 7.1 Phase Plane Trajectories of a Stick-Slip Cycle and the Corresponding Unperturbed Trajectory.

The trajectory of an undamped, second order system will be used as the unperturbed trajectory. Thus, in dimensionless coordinates the unperturbed trajectory gives a circle on the phase plane. The trajectory of an example stick-slip cycle and the corresponding unperturbed trajectory are shown in figure 7.1. The example true trajectory was generated by numerically integrating equation (7.12) with the parameters $\lambda_\infty^* = 2.0$; $\gamma^* = 5.9$; $\tau_L^* = 1.12$; $\dot{\xi}_d = 0.5$ and $\rho = 0.35$. The circle is the corresponding unperturbed trajectory. This example was selected to lie near the boundary at which stick-slip is eliminated.

Using the unperturbed trajectory the influence of damping and the friction forces can be calculated as perturbations. The problem is converted in this way from one of analyzing a differential equation, (7.12), into one of integrating a set of equations along a fixed path. Stick-slip can be detected as a condition on the sum of the perturbations. The usefulness of this analysis rests on the fact that, for a broad range of system parameters, the mass-spring characteristics dominate the motion of the system. This is particularly true near the stability boundary, where the perturbations sum to zero. Additionally, experimental evidence supporting this choice of unperturbed trajectory may be found in [Armstrong-Helouvry 89, 90], where it is observed that over a factor of 20 range in stiffness and 30 range in velocity, the slip distance of a stick-slip cycle of the PUMA is closely approximated by $x_{\text{slip}} = 2\pi\omega \, \dot{x}_d$; that is, the period of the unperturbed system times the desired velocity.

Figure 7.2, which presents a stick-slip trajectory on the phase plane, illustrates the definitions of terms used in this analysis. The axes of the plane are dimensionless position, ξ, and velocity, $\dot{\xi}$. Terms referring to breakaway are subscripted 'b', thus λ_b is the level of friction at breakaway, and λ_{b_n} is the level of friction at breakaway of the nth stick-slip cycle. The position at the moment of breakaway is ξ_b; and r_b is the radius of the trajectory at breakaway. The energy of the system at break-away is given by $E^* = \frac{1}{2}r_b^2$. Because $\dot{\xi} = 0$ describes the system moving at the desired velocity, $\dot{\xi} = -\dot{\xi}_d$ corresponds to the system with a physical velocity of zero, i.e., in the stuck condition. The time t_1^* is the (dimensionless) time spent during the slip portion of the cycle, and t_2^* is the time of the stick portion of the cycle.

We are concerned with the change in kinetic energy during an orbit on the phase plane. The dimensionless kinetic energy at break-away is $E_k^* = 1/2\dot{\xi}_d^{\,2}$. To interrupt the limit-cycle of stick-slip, the velocity after an orbit on the phase plane must not reach $\dot{\xi} = -\dot{\xi}_d$ ($\dot{x} = 0$ in physical coordinates). This is equivalent to the condition that the kinetic energy of the system be reduced during an orbit on the phase plane. Only systems

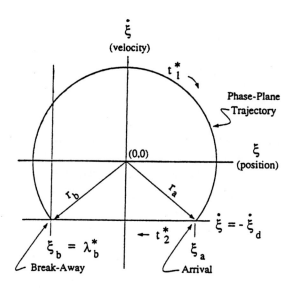

Figure 7.2   Phase Plane Trajectories Indicating the Definitions of Terms.

which exhibit steady motion or bounded frictional instability are considered here. Neglected are systems with divergent modes of instability, such as k_p negative. All of which is to say that it is assumed that either the system converges to its desired velocity or an orbit exits on the phase plane.

The change during an orbit of the dimensionless kinetic energy may be written:

$$\Delta_{E_k^*} = \int_{t_b^*}^{t_a^*} \ddot{\xi}\,\dot{\xi}\,dt^* \ . \tag{7.16}$$

Using (7.11) to replace $\ddot{\xi}$ gives:

$$\Delta_{E_k^*} = -\int_{t_b^*}^{t_a^*} (\xi)\,\dot{\xi}\,dt^* - \int_{t_b^*}^{t_a^*} 2\rho\dot{\xi}\,\dot{\xi}\,dt^* - \int_{t_b^*}^{t_a^*} \lambda^*\left(\gamma^*,t_2^*\right) f_s\left(\left(\dot{\xi}+\dot{\xi}_d\right),\tau_L^*\right)\dot{\xi}\,dt^* \ ;$$

$$\Delta_{E_k^*} = \Delta_{E_w^*} + \Delta_{E_\rho^*} + \Delta_{E_{f_s}^*} \ ; \tag{7.17}$$

where $\Delta_{E_k^*}$ is the change in dimensionless kinetic energy; and $\Delta_{E_w^*}$, $\Delta_{E_\rho^*}$ and $\Delta_{E_{f_s}^*}$ are symbols for the respective integrals. $\Delta_{E_w^*}$ is the energy contributed to the system through windup of the spring prior to break-away.

$\Delta_{E_p^*}$ is the energy dissipated by the damping term. And $\Delta_{E_{f_s}^*}$ is the energy contributed or dissipated by the integrated product of Stribeck friction and velocity. Exact evaluation of (7.16) would require knowledge of the state along the trajectory, which is to say the solution of (7.12). As a calculable approximation, equation (7.16) can be integrated along the path of the unperturbed trajectory.

The specifications of the unperturbed trajectory are:

$$\xi(t^*) = -\lambda_b^* f_s\left(\dot{\xi}_d, 0\right) - \dot{\xi}_d \sin(t^*); \quad \dot{\xi}(t^*) = -\dot{\xi}_d \cos(t^*); \quad 0 \le t^* \le 2\pi;$$
(7.18)

where $\dot{\xi}_d$ is the radius of the circle on the phase plane and the term $\lambda_b^* f_s\left(\dot{\xi}_d, 0\right)$ accounts for the fact that the equilibrium position is shifted by Stribeck friction when the desired velocity is low.

Because spring force is conservative, $\Delta_{E_w^*}$ may be written in terms of the beginning and final positions:

$$\Delta_{E_w^*} = \frac{1}{2}\xi_b^2 - \frac{1}{2}\xi_a^2 = \frac{1}{2}\lambda_b^{*2}\left(1 - \left(\frac{1}{1 + \dot{\xi}_d^2}\right)^2\right)$$
(7.19)

where $\xi_b^2 = \lambda_b^{*2}$, the point at which spring force exceeds static friction;
$\xi_a^2 = \lambda_b^{*2} f_s^2(\dot{\xi}_d, 0)$, the end point of the unperturbed trajectory.

Using the trajectory of (7.18) in the second integral of equation (7.17) gives the energy dissipated by damping:

$$\Delta_{E_p^*} = -\int_0^{2\pi} 2\rho\,\dot{\xi}\dot{\xi}\,dt^* = -2\pi\rho\,\dot{\xi}_d^2 .$$
(7.20)

The term $\Delta_{E_{f_s}^*}$ is the increase in system energy due to non-linear friction. Combining in the integral of equation (7.17), the friction model of (7.12) and the unperturbed trajectory of (7.18) gives:

$$\Delta_{E_{f_s}^*} = -\lambda_b^*\left[\int_0^{\tau_L^*}\dot{\xi}(t^*)\,dt + \int_{\tau_L^*}^{2\pi}\frac{1}{\left(\dot{\xi}(t^* - \tau_L^*) + \dot{\xi}_d\right)^2 + 1}\dot{\xi}(t^*)\,dt^*\right]$$

$$= -\lambda_b^*\dot{\xi}_d\left((-\sin(\tau_L^*)) + \int_{\tau_L^*}^{2\pi}\frac{1}{\dot{\xi}_d^2(1 - \cos(t^* - \tau_L^*))^2 + 1}(-\cos(t^*))\,dt^*\right)$$
(7.21)

Separating the integral from the multiplicative factors yields:

$$\Delta_{E_{j_\bullet}^\cdot} = \lambda_b^* \dot{\xi}_d \Delta_{E_{j_\bullet}^\cdot}' \quad ; \tag{7.22}$$

where

$$\Delta_{E_{j_\bullet}^\cdot}' = \sin\left(\tau_L^*\right) + \int_{\tau_L^*}^{2\pi} \frac{\cos\left(t^*\right)}{\dot{\xi}_d^2 \left(1 - \cos\left(t^* - \tau_L^*\right)\right)^2 + 1} \, dt^* \quad . \tag{7.23}$$

The integral of (7.23) can not be evaluated analytically, but can be evaluated numerically as a function of its two independent parameters, $\dot{\xi}_d$ and τ_L^*. The result of this numerical evaluation is shown in figure 7.3, where radial perturbation is plotted as a function of $\dot{\xi}_d$ and τ_L^*. Figure 7.3 confirms that as the frictional lag grows larger the destabilizing influence of the non-linear friction is reduced.

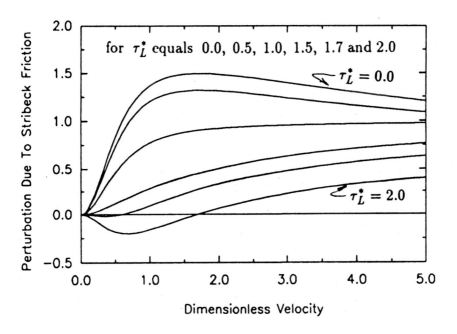

Figure 7.3 The Integral of Equation (7.23) Evaluated Numerically as a Function of $\dot{\xi}_d$ and τ_L^*.

Combining the terms of equation (7.17) gives the change in kinetic energy during an orbit on the phase plane:

$$\Delta_{E_k^*} \simeq \frac{1}{2} \lambda_b^{*2} \left(1 - \left(\frac{1}{1+\dot{\xi}_d^2}\right)^2\right) - 2\pi\rho\,\dot{\xi}_d^2 + \lambda_b^* \dot{\xi}_d \Delta_{E_{f_\bullet}'} . \qquad (7.24)$$

For $\rho > 0$, the second term makes a strictly negative contribution. With increasing stiffness λ_b^* decreases, decreasing the potentially positive contributions of the first and third terms. τ_L^* increases with increasing stiffness, reducing and finally making negative $\Delta_{E_{f_\bullet}'}$. Holding other terms constant and considering only the impact of system stiffness, equation (7.24) expressed in physical coordinates becomes:

$$\Delta_{E_k^*} = \frac{1}{k_p} a - \frac{1}{\sqrt{k_p}} b + \frac{1}{\sqrt{k_p}} c \qquad (7.25)$$

where

$$a = \frac{1}{2}\left(\frac{\lambda_b/M}{\dot{x}_s}\right)^2 \left(1 - \left(\frac{1}{1+(\dot{x}_d/\dot{x}_s)^2}\right)^2\right)$$

$$b = \pi k_v' \sqrt{M}\,(\dot{x}_d/\dot{x}_s)^2$$

$$c = \frac{\lambda_b/M}{\dot{x}_s}\dot{\xi}_d \Delta_{E_{f_\bullet}'}.$$

Above some value of stiffness, $\Delta_{E_k^*}$ will be negative and stick-slip is eliminated. Note that the second and third terms are both proportional to $1/\sqrt{k_p}$. In the absence of frictional lag, $\Delta_{E_{f_\bullet}'}$ and c would be independent of stiffness and, when $c > b$, (7.25) would give a positive $\Delta_{E_k^*}$ for any value of stiffness. Even for arbitrarily stiff systems such a model would predict stick-slip. In the presence of frictional lag, equation (7.25) is assured to go negative with increasing stiffness because c decreases and finally becomes negative, as indicated by figure 7.3. Frictional lag explains the observed fact that stick-slip can be eliminated by stiffening a system, [Rabinowicz 65]. For low velocities in systems with moderate friction, the level at which stick-slip will be eliminated corresponds roughly to:

$$\tau_L^* \geq \pi \qquad \text{or} \qquad k_p \geq M\frac{\pi^2}{\tau_L^2} . \qquad (7.26)$$

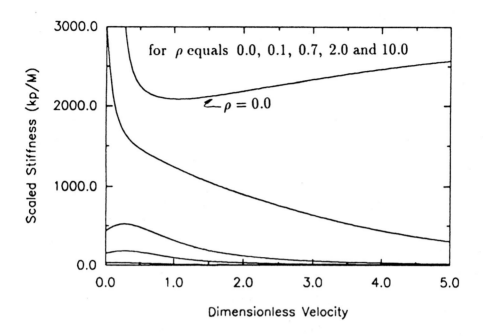

Figure 7.4 Stick-Slip Extinction Boundary as a Function of Normalized Stiffness and Dimensionless Desired Velocity. Calculated Using Equation (7.24).

The parameters of joint 1 of the PUMA arm are: $M = 6.3$ [kg-m^2], $\lambda_\infty = 1.6$ [kg-m/s^2], $\tau_L = 0.050$ [sec], $\gamma = 0.264$ [sec], $\dot{x}_s = 0.0058$ [rad/sec]. The viscous damping due to friction, measured during moderate to high velocity motions, is 4.94 [N-m/rad/sec]. Figure 7.4 was obtained by evaluating (7.24) using the parameters of the PUMA joint 1. Illustrated in figure 7.4 is the stick-slip extinction boundary as a function of normalized stiffness, k_p/M, and dimensionless desired velocity, $\dot{\xi}_d$. The extinction boundary is shown for several values of damping, ρ. The region below a stick-slip extinction boundary comprises combinations of system stiffness and desired velocity for which stick-slip will occur, as indicated by equation (7.24) and the condition $\Delta_{E_k^*} > 0$. In the region above an extinction boundary $\Delta_{E_k^*} < 0$ and stick-slip will not occur. As expected, the regime of stick-slip is reduced by increasing ρ. More intriguing is the fact that the required damping is a decreasing function of stiffness, and, perhaps unexpectedly, there are values of stiffness and desired velocity for which stick-slip will not occur, even if the system possess no damping. Systems which exhibit smooth motion in the absence of damping are possible because of the influence of frictional lag. The unusual mix of physical and dimensionless coordinates in figure 7.4, as well as figures 7.6, 7.7, 7.8 and 7.10, was chosen to enhance interpretation. In the dimensional analysis, stiffness is scaled to unity; and it is friction, $\lambda^*(\gamma^*, t_2^*)$, which scales inversely with physical stiffness and could be used as the dimensionless ordinate variable. But stiffness rather than friction is normally a variable under engineering control, and correspondingly, there is a developed intuition to aid interpretation. When $\tau_L = 0.050$ [sec], $k_p/M = 1974.0$ [rad/sec^{-2}] corresponds to $\tau_L^* = \pi$; this is approximately the value of normalized stiffness at which the undamped system of figure 7.4 is stabilized for low velocity motions.

Interpretation of figure 7.4 in the region of high stiffness and extremely low velocity must be made with some care. In this region the energy contributions of the three terms of equation (7.24) become exceeding small. As stiffness rises, the energy contributions to stick-slip diminish, as seen in equation (7.25). Likewise, as desired velocity decreases, the energy contributions diminish. Examination of equations (7.24) and (7.23) shows that as $\dot{\xi}_d \to 0$, $\Delta_{E_w^*} \to 0$ as $\dot{\xi}_d^2$, $\Delta_{E_p^*} \to 0$ as $\dot{\xi}_d^2$, and $\Delta_{E_{f_*}} \to 0$ as $\dot{\xi}_d^3$. Because $\Delta_{E_{f_*}} \to 0$ more rapidly than $\Delta_{E_w^*}$, equation (7.24) will predict that very high stiffness is required to extinguish stick-slip when ρ is small and velocity is very low. But each of the three energy contributions becomes extremely small, suggesting that if there is any unmodeled energy dissipation, the stiff, low-velocity system will not exhibit stick-slip.

7.3 The Impact of Static Friction Rising as a Function of Dwell Time

The level of static friction at breakaway, λ_b, is not constant. During motion the level of friction decreases. When motion stops, the static friction builds with time from a reduced value to its ultimate steady state value. This process is illustrated in figures 2.17 and 6.22. The process, outlined in section 2.3, is one of the boundary lubricant creeping out of the asperity junction sites. The implications of rising static friction for stick-slip are substantial: if the length of dwell time, t_2, is short in relation to the static friction rise time, γ, the static friction at breakaway is reduced and the amplitude of the stick-slip cycle diminished. This is illustrated in figures 2.15 and 2.18. As the amplitude of the stick-slip cycle is diminished, the dwell time is reduced and thus the level of static friction is further reduced. If an equilibrium is not reached in this cycle of diminishing static friction and dwell, stick-slip will be extinguished.

The interaction sliding friction and the rise of static friction is not fully understood. One issue that is uncertain, but must be quantified to support analysis, is the level of friction from which the rise of static friction begins, λ_a. Derjaguin *et. al.* , [57], employ a static plus kinetic friction model and presume that the rise of static friction commences from the level of kinetic friction. Rabinowicz, [58], suggests that at the instant of arriving at zero velocity, indicated by t_a or $t_2 = 0$, the increase of static friction over kinetic friction will be zero. If we consider a more complicated sliding friction model, one governed by the Stribeck curve and delay, a more complicated friction condition exists at t_a because the level of sliding friction rises as the system decelerates. Rising sliding friciton is shown by the data of Hess and Soom, [90] (see figures 2.21 and 2.22). The data from several sources, [Sampson *et. al.* 43; Vinogradov *et. al.* 67; Bell and Burdekin 69; Khitrik and Shmakov 87; Hess and Soom 90], suggest that the sliding friction rises to the level of static friction as *steady state* velocity approaches zero.

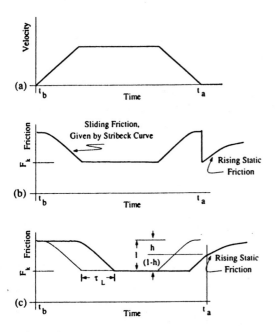

Figure 7.5 Two Possible Starting Points for the Rise of Static Friction.

In figure 7.5 a hypothetical velocity profile and corresponding friction profiles are shown. Curve 7.5(a) is that of the velocity profile and curves 7.5(b) and 7.5(c) are friction profiles of two possible friction models. Curves 7.5(b) and 7.5(c) differ most distinctly at time t_a. The friction profile of curve 7.5(b) is that given by the Stribeck friction curve and no frictional lag ($\tau_L = 0$). If the static friction were to start its rise from the level of kinetic friction, as suggested in [Derjaguin *et. al.* 57], a sudden drop in friction at t_a would be required. An alternative model provides that the rise of static friction begins from the level of kinetic friction at t_a, as shown schematically in figure 7.5(c). The level of kinetic friction at t_a, will be less than the value of static friction because of the interaction of system deceleration, frictional lag and Stribeck friction. Such a friction model is consistent with the dynamic friction model, allows sliding friction to go to the static friction value as steady-state velocity goes to zero, and provides that friction is not discontinuous at the instant t_a. Using this model, the interaction of λ, γ and τ_L will be studied. This portion of the analysis is not facilitated by the dimensional analysis and will be carried out in physical coordinates.

To represent the degree to which friction is reduced at time t_a the variable h is introduced:

$$\lambda_{a_n} = (1 - h)\, \lambda_{b_n} \; . \tag{7.27}$$

The variable h ranges from zero to one; zero indicating that there is no reduction in static friction during the slip portion of a stick-slip cycle, and one indicating that the level of static friction at the end of the slip cycle is reduced completely to the level of kinetic friction. Using the friction model, equation (7.2), and the unperturbed trajectory of the previous section, h can be evaluated:

$$h = \frac{\dot{\xi}_d^2\,(1 - \cos(\tau_L^*))^2}{\dot{\xi}_d^2\,(1 - \cos(\tau_L^*))^2 + 1} = \frac{\dot{x}_d^2\,(1 - \cos(\omega\tau_L))^2}{\dot{x}_d^2\,(1 - \cos(\omega\tau_L))^2 + \dot{x}_s^2} \; ;$$

$$h \approx \frac{1}{4}\left(\frac{\dot{x}_d}{\dot{x}_s}\right)^2 (\omega\tau_L)^4 \; . \tag{7.28}$$

The approximation is the first term of the Taylor series expansion and is valid when h is small. When the approximation would give a value of h larger than one, $h = 1$ may be used: h can not be greater than one. Combining the approximation and upper limit on h yields:

$$h \approx \begin{cases} \frac{1}{4}\left(\frac{\dot{x}_d^2}{\dot{x}_s^2}\right)^2 (\omega\tau_L)^4 \,, & k_p \dot{x}_d < \frac{2M\dot{x}_s}{\tau_L^2} \\[2ex] 1 \,, & k_p \dot{x}_d \geq \frac{2M\dot{x}_s}{\tau_L^2} \end{cases} \tag{7.29}$$

Combining the definition of h, equation (7.27), and the static friction model, equation (7.3), gives a discrete recursive equation for the level of static friction at the moment of breakaway, t_{b_n}:

$$\lambda_{b_n} = (1 - h)\,\lambda_{b_{n-1}} + \left(\lambda_\infty - (1 - h)\,\lambda_{b_{n-1}}\right)\frac{t_2}{t_2 + \gamma}$$

$$= \frac{\lambda_\infty t_2 + \gamma\,(1 - h)\,\lambda_{b_{n-1}}}{t_2 + \gamma} \,. \tag{7.30}$$

Equation (7.30) is more complex than it appears, because t_2 depends upon λ_b. A larger λ_b contributes to a larger $\Delta_{E_w^\bullet}$ and $\Delta_{E_{f_\bullet}^\bullet}$; both of which contribute to a larger r_a, resulting in a longer distance to travel during the stick portion of the cycle and thus a longer dwell time. In this way the cycle c-d-e of figure 2.15 has a longer dwell time than cycle i-j-k. The dependence of t_2 upon λ_b can be approximated by:

$$t_{2_n} = \frac{x_{a_{n-1}} - x_{b_n}}{\dot{x}_d} \approx \frac{2\,\lambda_{b_{n-1}}/k_p}{\dot{x}_d} \,. \tag{7.31}$$

Substituting (7.31) into (7.30) gives:

$$\lambda_{b_n} = \frac{\lambda_{b_{n-1}}\left(\lambda_\infty + \frac{1}{2}(1 - h)\,k_p\,\dot{x}_d\,\gamma\right)}{\lambda_{b_{n-1}} + \frac{1}{2}\,k_p\,\dot{x}_d\,\gamma} \,. \tag{7.32}$$

To facilitate analysis of the stability of (7.32), we write

$$\lambda_{b_n} = \frac{\lambda_{b_{n-1}}\,a}{\lambda_{b_{n-1}} + b} \,; \tag{7.33}$$

where

$$a = \left(\lambda_\infty + \frac{1}{2}(1 - h)\,k_p\,\dot{x}_d\,\gamma\right) \,;$$

$$b = \frac{1}{2}\,k_p\,\dot{x}_d\,\gamma \,;$$

$$b, \lambda_b > 0 \,.$$

The parameter b is evidently greater than zero (the sgn($\cdot$) functions of equation (7.1) handle the case that one might choose $\dot{x}_d < 0$). The existence of a stick-slip cycle is a necessary condition for the validity of update equation (7.32), giving rise to the requirement that $\lambda_b > 0$.

The map $\lambda_{b_{n-1}} \rightarrow \lambda_{b_n}$ has two fixed points: $\lambda_{b_{n-1}} = 0$ and $\lambda_{b_{n-1}} = (a-b)$. When $a > b$, the fixed point $\lambda_{b_{n-1}} = 0$ is a repulsor, that is it is unstable; and the fixed point $\lambda_{b_{n-1}} = (a - b)$ is an attractor, a stable fixed point. The update of (7.33) is nonlinear, but near the steady state it will converge with rate $((a - b)/a)^n$, where n is the index on stick-slip cycles. Away from the steady state the convergence rate is faster. Expressing the fixed point $\lambda_{b_{n-1}} = (a - b)$ in the parameters of the friction model yields $\lambda_{b_{**}}$, the steady state value of λ_b:

$$\lambda_{b_{**}} = \lambda_\infty - \frac{1}{2} h \, k_p \, \dot{x}_d \, \gamma \ . \tag{7.34}$$

When $(a - b) \leq 0$ there is no fixed point greater than zero. With rate not less than $(a/b)^n$, λ_{b_n} will decay to zero. As λ_{b_n} decays to zero, stick-slip is extinguished. Expressing the requirement $(a - b) > 0$ in terms of the parameters of the friction model yields the necessary condition for stick-slip:

$$\lambda_\infty > \frac{1}{2} h \, k_p \, \dot{x}_d \, \gamma \ . \tag{7.35}$$

When $\lambda_\infty \leq \frac{1}{2} h \, k \, \dot{x}_d \, \gamma$ there can be no stable stick-slip limit cycle. A system operating in this condition may experience several stick-slip cycles, simulations show this to happen, but the level of static friction decays steadily. Ultimately any non-zero damping is sufficient to extinguish the stick-slip cycle. Note that even though equation (7.34) can yield a negative value of the steady state λ_b, equation (7.32) will never actually give a negative value: λ_b decays until stick-slip is extinguished.

The parameter h may itself depend upon velocity and stiffness, as shown by equation (7.29). Considering this interaction, the stick-slip extinction boundary, the boundary in system parameter space on which $\lambda_{b_{**}} = 0$, is given by:

$$k_p \dot{x}_d \geq \begin{cases} \frac{2\lambda_\infty}{\gamma}, & \frac{2\lambda_\infty}{\gamma} > \frac{2M\dot{x}_*}{\tau_L^2} \\[2ex] 2 \sqrt[3]{\frac{\lambda_\infty}{\gamma} \frac{M^2 \dot{x}_*^2}{\tau_L^4}}, & \frac{2\lambda_\infty}{\gamma} \leq \frac{2M\dot{x}_*}{\tau_L^2} \end{cases} \tag{7.36}$$

The case statement arises from the cases of equation (7.29).

Plotted on the stiffness-velocity plane, equation (7.36) produces a curve $k_p \propto \frac{1}{\dot{x}_d}$; where the constant of proportionality is given by the dynamic and friction parameters of the system. In figure 7.6 the $\rho = 0$ curve is the contour given by equation (7.35), evaluated using the PUMA joint 1 parameters. This line is the extinction boundary due to the rise time of static friction when the system is undamped. As expected, figure 7.6 shows that smooth motion is more easily obtained as desired velocity is increased.

Using equation (7.34), the steady-state static-friction level may be determined. Incorporating $\lambda_{b_{**}}$ into equation (7.24) yields the change in

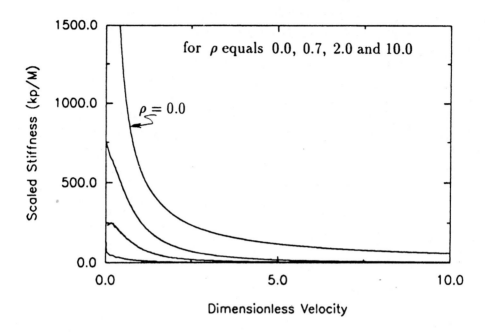

Figure 7.6 Stick-Slip Extinction Boundary as a Function of Stiffness and Desired Velocity. The $\rho = 0$ Contour was Calculated Using Equation (7.36); Other Contours were Calculated using Equation (7.37).

dimensionless kinetic energy during an orbit on the phase plane:

$$\lambda_{b_{..}} = \lambda_\infty - \frac{1}{2} h \, k_p \, \dot{x}_d \, \gamma \; ;$$

$$\lambda^*_{b_{..}} = \frac{\lambda_{b_{..}} / M}{\omega \, \dot{x}_s} \; ;$$

$$(7.37)$$

$$\Delta_{E^*_k} \simeq \frac{1}{2} \lambda^{*2}_{b_{..}} \left(1 - \left(\frac{1}{1 + \dot{\xi}_d^2} \right)^2 \right) - 2 \pi \rho \, \dot{\xi}_d^2 + \lambda^*_{b_{..}} \, \dot{\xi}_d \, \Delta'_{E^*_{j_.}} \; .$$

Equation (7.37) reflects the full friction model: non-linear Stribeck friction, frictional lag and rising static friction are considered. The contour on which $\Delta_{E^*_k} = 0$ is the stick-slip extinction boundary. When $\Delta_{E^*_k} < 0$ there will be no stick-slip because, after an orbit on the phase plane, the sliding velocity does not return to $\dot{x} = 0$. Equation (7.37) was used to calculate the $\rho = 0.7$, 2.0 and 10 contours of figure 7.6, and the $k_v = 4.9$ contour of figure 7.8.

Equation (7.37) provides the means to predict the presence of stick-slip using system and friction parameters. The analysis is approximate; the perturbation analysis, Taylor series expansion and approximation of t_2, as well as approximation in the underlying physical modeling, all contribute to the uncertainty of the result. The usefulness of the result for a specific range of parameters can be assessed by comparison with the results of numerical investigation. And, of course, when experiment is available, it can be used to test the validity of the approximate analytic result.

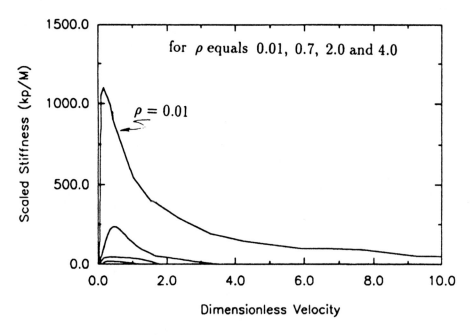

Figure 7.7 Stick-Slip Extinction Boundary as a Function of Stiffness and Desired Velocity; Determined by Numerical Integration of Equation (7.1).

Figure 7.7 was generated by numerical integration of equation (7.1). The stabilizing value of ρ was determined by iteration at each of 8,000 points. The iso-ρ contours were created by interpolation. Damping levels calculated with equation (7.37), presented in figure 7.6, are conservative, but otherwise reflect the simulation values quite well.

Perhaps more to the point, the results of 17 experimental trials are shown in figure 7.8, along with the extinction boundary calculated using equation (7.37) and the measured frictional damping of $F_v = 4.9$ [N-m/rad/sec]. The condition used to calculate the extinction boundary of figure 7.8 is $\Delta_{E_k^*} > \epsilon$,

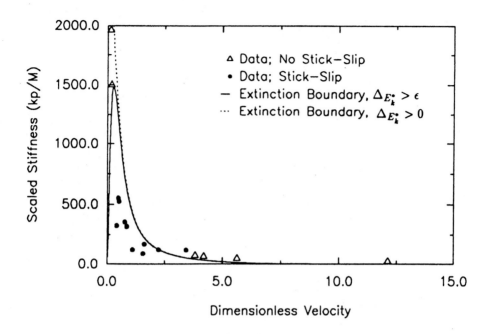

Figure 7.8 Experimental Stick-Slip Data; the Solid Contour is the Stick-Slip Extinction Boundary Calculated with the Condition $\Delta_{E_k^*} > \epsilon$; the Dotted Contour Corresponds to $\Delta_{E_k^*} > 0$.

rather than $\Delta_{E_k^*} > 0$. The parameter ϵ indicates the minimum excitation energy required to sustain stick-slip and has been empirically adjusted to the value of $\epsilon = 0.02$, or 2.1 micro-Joules per stick-slip cycle in physical coordinates, (see eqn. (7.14)). An energy dissipation of 2.1 micro-Joules per stick-slip cycle corresponds to roughly one 800th of the total frictional energy to be dissipated per cycle if stick-slip had occurred in the experimental trial run with $k_p/M = 1500$ and $\dot{\xi}_d = 0.185$. While ϵ has been empirically adjusted, it is not difficult to believe that this amount of enery could be dissipated by a process that has not yet been considered. As seen in figure 7.8, the empirical term makes a perceptible contribution only at high stiffness and low velocity.

Of the 17 experimental trials in figure 7.8, 6 show smooth motion. All of these lie above the stick-slip extinction boundary. Of the 11 trials that show stick-slip (two trials lie near to each other at $\dot{\xi}_d = 0.85, k_p/M = 320$), 10 would be predicted to stick-slip by this analysis. The 11th appears to be an outlier. With the exception of ϵ, the evaluation of the theoretical model is based upon the directly measured friction parameters presented in chapter

6; there is no parametric fitting of the friction model to the locations of the experimental points shown in figure 7.8.

7.4 Integral Control

The system considered thus far has incorporated only proportional-derivative (PD) control. Integral control is often applied to systems with friction to reduce and sometimes eliminate steady state errors. The impact of integral control on stick-slip, however, may be quite complex. One common integral control implementation, proportional-integral-derivative (PID) control, will be considered here.

PID control allows placement of a pole to boost low frequency gain. The closed loop pole is normally well within the bandwidth of the system. Writing y in a form slightly different from that of equation (7.4),

$$y = x - x_d \ , \tag{7.38}$$

the control law may be written:

$$u = -k_i \int y \, dt \ - k_p \, y \ - k_v \, \dot{y} \ ; \tag{7.39}$$

giving the system dynamic equation:

$$M \, \ddot{y} = -k_i \int y \, dt \ - k_p \, y \ - k_v' \, \dot{y} \, \lambda \, (\gamma, t_2) \, f_s \, (\cdot) \ - F_k \, \text{sgn} \, (\dot{y}) \ . \tag{7.40}$$

Developing the dimensionless model as before, and using the fact that

$$\int y \, dt \ = \ \frac{\alpha}{\beta} \int \xi \, dt^* \ , \tag{7.41}$$

gives the dimensionless dynamic model:

$$\ddot{\xi} = \ -k_i^* \int \xi \, dt^* - \xi \ - \ 2\rho \dot{\xi} \ - \ \lambda^* \, (\gamma^*, t_2^*) \, f_s \, (\cdot) \ - \ F_k^* \ ;$$

where $k_i^* = (k_i/M) \, /\beta^3$. In these coordinates, k_i^* is approximately equal to the frequency of the integral compensation pole as a fraction of the natural frequency of the system. Values of k_i^* in the range 0.05 - 0.2 would be common settings for a control system gain†.

To apply perturbation analysis to the system with integral control it is important to choose an unperturbed trajectory that will give a good approximation to the integral of position error. The unperturbed trajectory used here is made up of a portion reflecting the path during slip and a portion reflecting the path during stick. Recalling that ξ_b is the position at the moment of break-away and is negative‡; the path during slip, $0 < t^* < 2\pi$, is given by:

$$\dot{\xi}(t^*) = -\frac{2\xi_b}{2\pi} - \dot{\xi}_d \cos(t^*) \quad ;$$

$$\xi(t^*) = -\frac{2\xi_b}{2\pi} t^* + \xi_b - \dot{\xi}_d \sin(t^*) \quad ; \qquad (7.43)$$

$$\int \xi \, dt^* = -\frac{\xi_b}{2\pi} t^{*2} + \xi_b t^* + \dot{\xi}_d \cos(t^*) + C_F + C_{\bar{0}} \quad ;$$

where C_F and $C_{\bar{0}}$ are the constant of integration broken into two parts. C_F offsets the kinetic and steady state viscous friction: $C_F = -\left(F_k^* + 2\rho\dot{\xi}_d\right)/k_i^*$. $C_{\bar{0}}$ is chosen so that the mean of the integral control action over a stick-slip cycle is zero.

---

† To first approximation, the shift in pole location resulting from the addition of the integral term can be seen by writing the characteristic equation in the form: $s\left(s^2 + k_v s + k_p\right) = -k_i$; k_p, k_v, k_i scaled by M. Expanding s into $(s_0 + \epsilon)$, where s_0 represents the roots when $k_i = 0$, i.e., the roots of the second order equation plus $s_0 = 0$; and solving for ϵ, keeping only the first order terms, gives: $\epsilon = -k_i/k_p$; $k_i/(2k_p + k_v s_0)$. The pole shifted by $\epsilon = -k_i/k_p$ is the one referred to above. Translated to dimensionless coordinates, $\epsilon/\omega = k_i^*$. The shift of the second order poles, $\epsilon = k_i/(2k_p + k_v S_0)$, is a shift toward the stability boundary of approximately $k_i^*/2$ in dimensionless coordinates, indicating the degree to which the integral term diminishes the damping corresponding to k_p and k_v. This approximation is valid over a considerable range of k_i when k_v is moderate, that is, when the roots of the second order system are complex.

‡ When k_i^* is sufficiently large and $\dot{\xi}_d$ sufficiently small, break-away at a position $\xi_b > 0$ is possible. This event would result in direction reversal and marks the onset of hunting, a limit cycle arising from the integral control when $\dot{\xi}_d \approx 0$. Hunting is not considered here.

The path during the stick portion of a stick-slip cycle, $2\pi < t^* < 2\pi + t_2^*$, is given by:

$$\dot{\xi}(t^*) = -\dot{\xi}_d \; ;$$

$$\xi(t^*) = -\dot{\xi}_d t^* + \dot{\xi}_d \left(2\pi + \frac{1}{2}t_2^*\right) \; ;$$

$$\int \xi \, dt^* = -\frac{1}{2}\dot{\xi}_d t^{*2} + \dot{\xi}_d \left(2\pi + \frac{1}{2}t_2^*\right) t^*$$

$$+ \left(C_F + C_{\bar{0}} + \dot{\xi}_d - \dot{\xi}_d \left(2\pi^2 + t_2^*\pi\right)\right) \; . \tag{7.44}$$

The constant of integration, $\left(C_F + C_{\bar{0}} + \dot{\xi}_d - \dot{\xi}_d \left(2\pi^2 + t_2^*\pi\right)\right)$, has been chosen to match (7.44) with (7.43) at $t^* = 0$ and $t^* = 2\pi$.

Canceling C_F with the kinetic and steady viscous friction forces, the mean integral control action is given by:

$$\int_0^{2\pi+t_2^*} \int_0^{t^*} \xi(\tau) \, d\tau \, dt^* = \frac{2}{3}\pi^2 \xi_b + \dot{\xi}_d \left(\frac{t_2^{*3}}{12} + t_2^*\right) + C_{\bar{0}}\left(2\pi + t_2^*\right) \; . \tag{7.45}$$

For the unperturbed trajectory the dwell time is given by $t_2^* = -2\xi_b/\dot{\xi}_d$. Incorporating t_2^* and solving for $C_{\bar{0}}$ by setting the mean integral control action to zero gives:

$$C_{\bar{0}} = \frac{-\xi_b \left(\frac{2}{3}\pi^2 - 2 - \frac{2}{3}\frac{\xi_b^2}{\dot{\xi}_d^2}\right)}{2\pi - 2\xi_b/\dot{\xi}_d} \; . \tag{7.46}$$

$C_{\bar{0}}$ ensures that the net force of the integral control term is zero, or, considering C_F, just sufficient to cancel the steady state sliding friction force. The integral control action at the moment of break-away, however, need not be zero.

Defining F_{ib}^* to be the force applied at break-away by the integral control term,

$$F_{ib}^* = -k_i^* \left(C_{\bar{0}} + \dot{\xi}_d\right) \; , \tag{7.47}$$

the force balance at break-away is given by:

$$\lambda_b^* = -\xi_b + F_{ib}^* \; ; \tag{7.48}$$

noting that in dimensionless coordinates $k_p^* = 1$, thus the force due to the proportional term is ξ_b. Expressing the contribution of integral control to windup as a fraction of the total windup, ξ_b, one finds:

$$\frac{F_{ib}^*}{-\xi_b} = -k_i^* \left(\frac{\left(\frac{2}{3}\pi^2 - 2 - \frac{2}{3}\left(\frac{-\xi_b}{\xi_d} \right)^2 \right)}{2\pi + 2\left(\frac{-\xi_b}{\xi_d} \right)} + \left(\frac{-\xi_b}{\dot{\xi}_d} \right)^{-1} \right) ; \qquad (7.49)$$

where the minus sign is carried with ξ_b because ξ_b is a negative number. When $\frac{F_{ib}^*}{-\xi_b} = 0$, integral control has no effect on break-away: the integral of position error is passing through zero just at the moment of break-away. This occurs when $\left(\frac{-\xi_b}{\xi_d} \right) \simeq 3.5$, or $\lambda_b^*/\dot{\xi}_d \simeq 3.5$. When λ_b^* is large the integral control accelerates break-away by reducing $|\xi_b|$. For $k_i^* = 0.15$, $|\xi_b|$ is halved when $\left(\frac{-\xi_b}{\xi_d} \right) = 23.2$, or $\lambda_b^*/\dot{\xi}_d = 46.4$. When λ_b^* is small, the integral control term acts to retard break-away by increasing $|\xi_b|$. For $k_i^* = 0.15$, $|\xi_b|$ is doubled when $\left(\frac{-\xi_b}{\xi_d} \right) = 0.375$, or $\lambda_b^*/\dot{\xi}_d = 0.188$. Considering only $\Delta_{E_w^*}$, accelerating break-away - $\lambda_b^*/\dot{\xi}_d$ larg - tends to reduce excitation of the stick-slip cycle. Likewise, retarding break-away - $\lambda_b^*/\dot{\xi}_d$ small - tends to increase excitation of the stick-slip cycle. Combining these effects with the influence of integral control during the slip portion of the cycle - to be calculated next - the overall impact of integral control can be determined.

As do the other forces of motion, the integral control term will make a contribution to the change in kinetic energy during a the slip portion of a stick-slip cycle, that is, a contribution to equation (7.17). The contribution may be evaluated using the unperturbed trajectory of equation (7.43):

$$\Delta_{E_i^*} = \int_0^{2\pi} -k_i^* \int_0^{t^*} \xi(\tau)\, d\tau\, \dot{\xi}(t^*)\, dt^* ;$$
$$\Delta_{E_i^*} = -k_i^* \left[\left(\frac{8\pi}{6} - 2 \right) \xi_b^2 + 2\xi_b\dot{\xi}_d - 2C_0^-\xi_b - \dot{\xi}_d^2 \pi \right] . \qquad (7.50)$$

$\Delta_{E_i^*}$ is a complex function of break-away position and desired velocity, made more so by the fact that C_0^- may be either signed and is itself a rational polynomial in these variables. Expanding C_0^- in equation (7.50) and factoring out a term ξ_d^2 yields a rational polynomial in $\left(\frac{-\xi_b}{\xi_d} \right)$ with a quartic numerator and a non-zero, first order denominator. The numerator has real roots at $\left(\frac{-\xi_b}{\xi_d} \right) = 1.45$ and 5.35. Between these values $\Delta_{E_i^*}$ is negative and the integral control term tends to diminish the kinetic energy during the slip cycle. For $\left(\frac{-\xi_b}{\xi_d} \right) < 1.45$ or $\left(\frac{-\xi_b}{\xi_d} \right) > 5.35$, $\Delta_{E_i^*}$ is positive and encourages stick-slip.

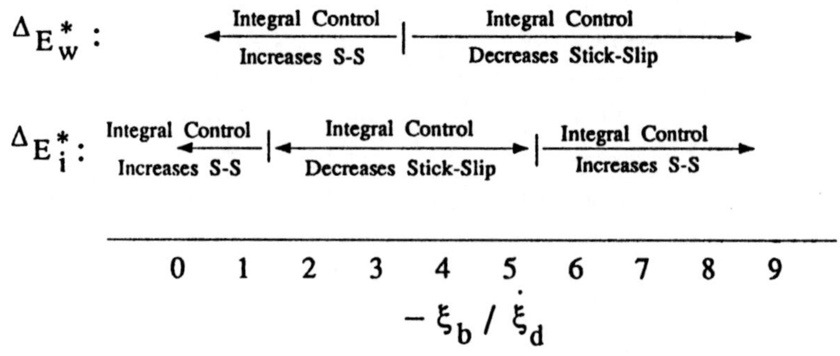

$\Delta_{E\,w}^{\,*}$:

Integral Control	Integral Control
Increases S-S	Decreases Stick-Slip

$\Delta_{E\,i}^{\,*}$:

Integral Control	Integral Control	Integral Control
Increases S-S	Decreases Stick-Slip	Increases S-S

$$0 \quad 1 \quad 2 \quad 3 \quad 4 \quad 5 \quad 6 \quad 7 \quad 8 \quad 9$$

$$-\xi_b\,/\,\dot{\xi}_d$$

Figure 7.9 Influence of integral control as a function of $\left(\frac{-\xi_b}{\dot{\xi}_d}\right)$.

The influence of integral control on the two perturbations contributing to stick-slip is presented in figure 7.9. The exact range of $\left(\frac{-\xi_b}{\dot{\xi}_d}\right)$ over which integral control reduces stick-slip will depend upon the relative importance of $\Delta_{E_w^*}$ and $\Delta_{E_i^*}$, and can be calculated using the parameters of a specific system. Exact calculation of the contribution of integral control to $\Delta_{E_i^*}$ can be carried out by evaluating ξ_b as implied by (7.48) and using the result in (7.19) to compute $\Delta_{E_w^*}$. $\Delta_{E_i^*}$ must also be computed. In many practical applications of integral control, such a result would have to be modified to account for the use of a deadband or saturation operation. A Deadband is used with integral control to reduce the possibility of hunting, the friction induced limit cycle when $\dot{x}_d = 0$. A saturation operation is used with integral control when large motion transients might saturate the actuators. In specific instances (7.50), (7.48) and (7.19) may be evaluated, perhaps with modification to reflect a deadband or saturation in the integral control.

Plotted in figure 7.10 are simulation results, similar to those of figure 7.7 except that integral control is implemented with $k_i^* = 0.15$, thereby setting the compensator pole at a frequency of roughly $1/6^{th}$ the system's natural frequency. For figure 7.10, ρ is given by $\rho = k_v/(2\omega)$, as it is elsewhere in this chapter; this does not reflect the pole location of the third order system and should be considered a scaled representation of k_v. The $\rho = 0.1$ contour in figure 7.10 corresponds to a damping factor of 0.025 for the complex poles of the third order system. This contour lies above the $\rho = 0.01$ contour in figure 7.7. Near the undamped extinction boundary of figure 7.6 the system with integral control requires considerably more damping to extinguish stick-slip than does the system without integral control. Over the entire graph the damping required to stabilize system with integral control is greater than

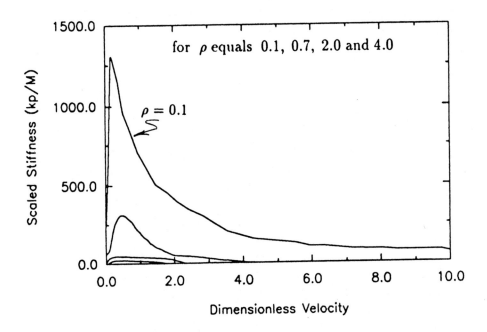

Figure 7.10 Stick-Slip Extinction Boundary as a Function of Stiffness and Desired Velocity; Determined by Numerical Integration of Equation (7.42).

that required to stabilize the second order system, exclusive of a small region near $k_p/M = 50$, $\dot{\xi}_d = 1.0$. Noting that λ_∞^* scales inversely with $\sqrt{k_p/M}$, the region of the lower left corner is, in dimensionless coordinates, the region of high friction and low velocity. In this region λ_b^* is essentially unmodified by dwell time: equation (7.34) gives a reduction in static friction of 0.01%. At $k_p/M = 50$ and $\dot{\xi}_d = 1.0$, λ_b^* is 5.90. Iteration on equations (7.46), (7.47) and (7.48) yields $\xi_b = -5.25$ and $\left(\frac{-\xi_b}{\dot{\xi}_d}\right) = 5.25$, a value for which one would expect reduced $\Delta_{E_w^*}$ and a negative (stabilizing) $\Delta_{E_i^*}$. For these parameters windup is reduced by integral control: equations (7.47) and (7.19) show a change in $\Delta_{E_w^*}$ from 13.05 to 10.33. For $\xi_b = -5.25$ and $\dot{\xi} = 1.0$, equation (7.46) yields $C_0^- = -4.32$, and equation (7.50) gives $\Delta_{E_i^*} = -0.20$. The total frictional excitation of stick-slip is given by:

$$\Delta_{E_k^*} = \Delta_{E_w^*} + \Delta_{E_p^*} + \Delta_{E_{j_s}^*} + \Delta_{E_i^*}$$

$$= 10.33 - 2\pi\rho\dot{\xi}_d + 7.38 - 0.20 \; ; \tag{7.51}$$

and a damping factor of $\rho = 2.80$ is required to extinguish stick-slip. Without integral control, (7.37) gives $\rho = 3.25$, confirming that integral control can reduce the required damping for this range of λ_b^* and $\dot{\xi}_d$.

Application to Observed Limits on Feedback Control

In section 6.2 feedback control experiments are described which give a lower velocity limit of 0.015 [rad/sec] for control based on integrated acceleration feedback, and a lower velocity limit of 0.15 [rad/sec] for the industrial controller. The friction curve of the mechanism, figure 6.18, shows two breaks. With respect to the first break, the velocity achieved by integrated acceleration feedback corresponds to a dimensionless velocity of $\dot{\xi}_d = 2.59$. Evaluating $\lambda_{b_{..}}$ using $k = 500$ [N-m/rad] gives $\lambda_{b_{..}} = 4.82$; and equation (7.24) gives a required damping to eliminate stick-slip of $\rho = 0.57$. The calculated figure of $\rho = 0.57$ is close to the controller damping of $\rho = 0.80$. Furthermore, figure 7.7 shows that required damping is a strong function of velocity in this velocity-stiffness regime. This analysis would have predicted a low velocity limit for the PUMA arm and integrated acceleration feedback that is quite close to that observed.

The industrial controller exhibits stick-slip beginning at the desired velocity of 0.15 [rad/sec]. The velocity is well above that of the first break at $\dot{x}_s = 0.0058$, but with respect to the characteristic velocity of the second break, $\dot{x}_s = 0.068$ [rad/sec], $\dot{x}_d = 0.15$ [rad/sec] gives a dimensionless velocity of $\dot{\xi}_d = 2.2$. The mechanism plus controller have a damped natural frequency of 2 Hz, suggesting $k_p = 950$ [N-m/rad]. Scaling λ_∞ yields $\lambda_\infty^* = 0.30$. Evaluating equation (7.24) results in a required damping of $\rho = 0.022$, certainly much less than the actual damping of the system. The low velocity friction is expected to exert a destabilizing influence, but not one adequate to account for the onset of stick-slip.

Explanation of the onset of stick-slip at a velocity of 0.15 [rad/sec] lies perhaps in the observation that at this velocity the motor pinion gear turns with a rate of 1.49 rev/sec, giving a pinion gear tooth rate of 17.9 teeth per second. A power spectrum of the arm responding to vibrational noise is presented in figure 6.23. Apparent in this figure is a large peak at 18 Hz; this peak corresponds to the first resonant mode of the arm plus transmission. It is perhaps the case that this mode is excited by the 2 [N-m], 18 Hertz friction signal arising from the teeth of the motor pinion gear (see figure 5.2, the signal which makes 6 cycles across the figure is at the spatial frequency of the motor pinion teeth). It is likely that the frictionally induced oscillations are sufficient to excite stick-slip, resulting in the low velocity limit of 0.15

[rad/sec]. Those in robotics who are investigating direct and homogeneous drive mechanisms may be on to more than they know.

Summary

Dimensional and perturbation analysis have been applied to the problem of non-linear low-velocity friction. Through dimensional analysis an exact model of the non-linear system can be formed in five parameters rather than ten, greatly facilitating study and explicitly revealing the interaction of parameters. By converting the system of differential equations. into a set of integrations, the perturbation technique makes approximate analysis possible where only numerical techniques had been available before. Comparison of figures 7.4 and 7.6 with figure 7.7 shows the approximation to be valid over a broad range of stiffness and velocity. Derived using the measured friction parameters of the PUMA arm, the predicted regions of stick-slip shown in figure 7.8 agree remarkably well with experimental observation.

Integral control will often encourage stick-slip. Only when $\left(-\xi_b/\dot{\xi}_d\right)$ lies in a moderate range, roughly from one to seven, will the integral term reduce $\Delta_{E_k^*}$ and thereby reduce stick-slip. When $\left(-\xi_b/\dot{\xi}_d\right)$ is small, the integral control is pushing when it should be pulling and delays break-away, contributing to $\Delta_{E_w^*}$. When $\left(-\xi_b/\dot{\xi}_d\right)$ is large, the integral control itself winds up and contributes a significant destabilizing force during slip. In physical coordinates $\left(-\xi_b/\dot{\xi}_d\right)$ is given by:

$$\left(\frac{-\xi_b}{\dot{\xi}_d}\right) = \frac{\lambda_b}{\dot{x}_d\sqrt{k_p m}} \ . \tag{7.52}$$

This the dimensionless group has been employed previously by [Derjaguin *et al.* , [57], as a measure of frictional excitation, and is a useful quantity in estimating the difficulty of eliminating stick-slip.

For joint 1 of the PUMA arm, $\left(-\xi_b/\dot{\xi}_d\right)$ is in the range where integral control will reduce stick-slip when $\dot{x}_d \sim 0.006$ [rad/sec] and $k_p/M \sim 100-600$ [rad/sec^2]. This region of the velocity-stiffness plane is of practical importance in the use of that mechanism. The range of conditions where integral control will reduce stick-slip may be compared with the range over which stick-slip is not already extinguished by frictional lag or rising static friction. For lighter systems, or those with smaller τ_L or γ, integral control will reduce stick-slip over a larger range of conditions. For heavier systems, or those with larger τ_L or γ, there may be no operating conditions in which integral control will reduce stick-slip.

Dimensional and perturbation analysis techniques have been used to study low-velocity friction and the most common control configurations. These techniques also provide a basis for the study of more specialized control schemes. Variable structure control (VSS for Variable Structure System) is, for example, strongly indicated by the need near zero velocity for control stiffness that may be impractical at larger velocities. Any of a number of VSS schemes may be studied within the dimensional and perturbation analysis framework by ensuring that the unperturbed trajectory captures the relevant features and performing the appropriate integrations. Alternatively, the analysis of stick-slip motion may be employed to determine the velocity below which stick-slip will occur. When this limit is known, control can be designed to operate in this regime with explicitly unsteady motions. In the next chapter, demonstrations of friction compensated motion are presented, including a demonstration of force control using the PUMA arm. The force control is executed by a controller designed to employ explicitly unsteady motions.

Chapter 8

Demonstrations of Friction Compensation

To demonstrate the accuracy of the friction model open-loop moves were undertaken. For the spatial motions demonstrated in sections 8.1 and 8.2, friction was predicted using a three component model:

1. Kinetic Friction ;
2. Viscous Friction ;
3. A Table Lookup to Compensate for Position-Dependent Friction.

Low-velocity non-linear friction was not compensated.

8.1 Open-Loop Motion of One Joint

Figure 8.1 is a plot of the desired and actual robot trajectory during a 20 second reciprocating motion. This motion was conducted open-loop: there is no feedback correction of errors. The motion of figure 8.1 is, in fact, the *first* reciprocating motion conducted. Friction compensation is based on a model and parameters developed as described in chapters 5 and 6; not on parameters identified on the trajectory shown. The applied torque is made up of the inertial and frictional contributions given by:

$$\tau = M\ddot{\theta}_d + F_k \operatorname{sgn}(\dot{\theta}_d) + F_v \dot{\theta}_d + F_{\mathrm{TL}}(\theta)$$

where $\ddot{\theta}_d, \dot{\theta}_d$ are desired acceleration and velocity;
M is the mass;
F_k is the kinetic friction;
F_v is the viscous friction;
F_{TL} is the table lookup compensation for position-dependent friction;
θ is the measured position.

The only state-dependent portion of the torque is the table lookup compensation for position-dependent friction. The parameters used are shown in table 8.1.

Table 8.1 Parameters used for the Motions of Figures 8.1 and 8.2.

	The Motion of Figure 8.1	The Motion of Figure 8.2	
Mass;	4.900	5.115	[Kg-m^2]
Viscous, positive rotation;	4.938	5.070	[N-m/rad/sec]
Viscous, negative rotation;	3.446	3.772	[N-m/rad/sec]
Kinetic, positive rotation;	8.432	8.121	[N-m]
Kinetic, negative rotation.	-8.262	-7.907	[N-m]

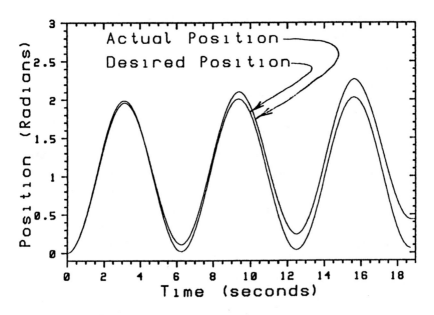

Figure 8.1 Desired and Actual Position Recorded During an Open-Loop Motion, Made using Parameters Determined by Direct Measurement.

As mentioned, the motion of figure 8.1 was generated with a parameter set general to the workspace. It is generally true that a set of parameters identified using data of a specific motion will afford better accuracy on that specific motion; the motion of figure 8.2 demonstrates this; the motion of figure 8.2 was conducted using parameters identified using data collected during the motion of figure 8.1. The final position error at the end of the motion of figure 8.2 is only 1.02 deg, or 0.14% of the total motion

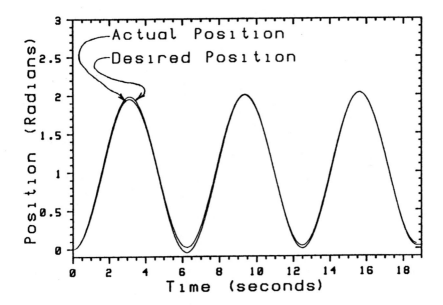

Figure 8.2   Desired and Actual Position Recorded During an Open-Loop Motion, Made Using Parameters Identified on the this Trajectory.

distance.   The motion of figure 8.1 provides, however, a better indication of the global accuracy of the open-loop friction compensation; this motion shows an accumulated position error of 4.1% - still not bad for open-loop control.

The total torque applied to bring about the motion of figure 8.2 is shown in figure 8.3. In figure 8.4 this torque is broken down into the inertial and friction components. The RMS inertial torque is 3.62 Newton-meters ([N-m]), the RMS friction torque is 10.72 [N-m], nearly 3 times greater. In figure 8.5 the friction torque is further broken down into the kinetic, viscous and table lookup torques.

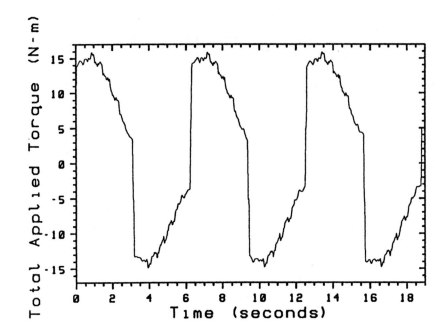

Figure 8.3   Total Torque Applied During the Motion of Figure 8.2.

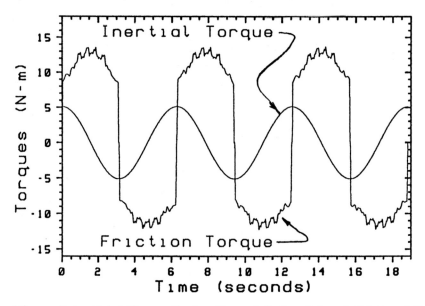

Figure 8.4   Total Torque Broken Down into Inertial and Frictional Torque
Components.   Note that the Inertial Torque is Never Great
Enough to Overcome Static Friction.

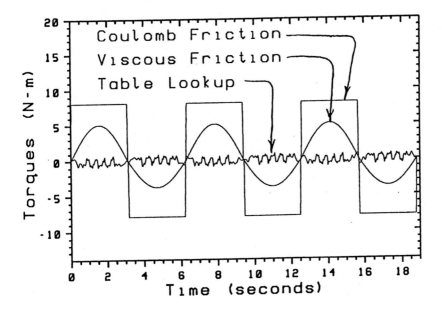

Figure 8.5   Friction Torque Broken Down into Kinetic, Viscous and Table Lookup Components.

8.2 Open-Loop Motion of Three Joints

To extend the friction model to joints 2 and 3, the break-away experiment of chapter 5 was carried out and a position-dependent friction compensation table developed. The constant-torque experiment of section 6.1 was carried out to determine the viscous and kinetic friction parameters. The positive direction position-dependent friction compensation tables developed for joints 2 and 3 are shown in figures 8.6 and 8.7. Roughly 60,000 measurements of static friction were made to build each table.

Data to measure friction as a function of velocity were collected at four torque levels for joint 2 and three torque levels for joint 3. At each torque level five measurements were made, allowing the variance to be estimated. These data are shown in figures 8.8 and 8.9.

Compensation for position dependencies in the friction data for joints 2 and 3 was not as successful as that for joint 1. Figure 8.10 shows a bang-coast-bang move of joint 3 comparable to those of figure 5.4. The RMS velocity error is 0.054 rad/sec in the motion of figure 8.10, vice 0.008 rad/sec in the motions of joint 1 shown in figure 5.4. One thing confounding the identification of friction at joints 2 and 3 is the gravity loading on these joints. During the breakv-away experiment the gravity load was compensated by a configuration-dependent calculation [Armstrong, Khatib and Burdick

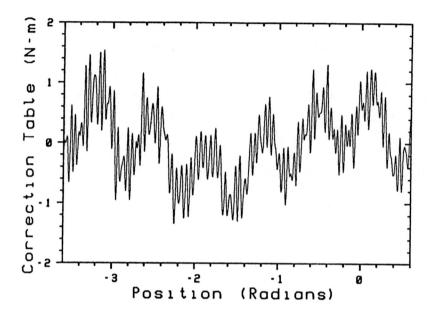

Figure 8.6   Lookup Table Correction to Torques Applied at Joint Two of the PUMA 560.

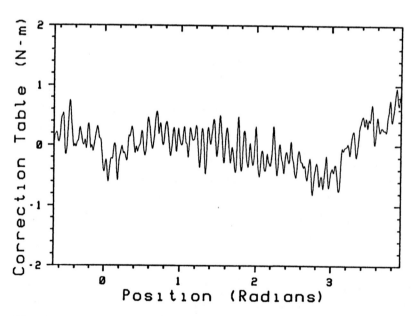

Figure 8.7   Lookup Table Correction to Torques Applied at Joint Three of the PUMA 560.

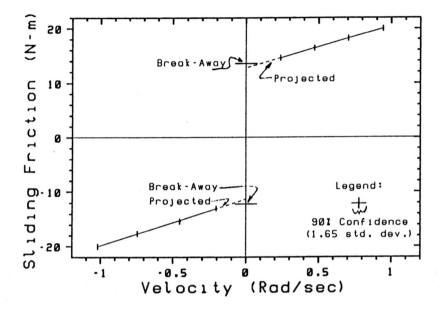

Negative Direction
Viscous 8.53   N-m/rad/sec
Kinetic 11.34 N-m

Positive Direction
Viscous 7.67   N-m/rad/sec
Kinetic 12.77 N-m

Figure 8.8.   Friction torque as a function of velocity, Joint 2.

86] that provided a gravity balancing torque added to the experimental torque. The RMS velocity error of 0.054 [rad/sec] corresponds roughly to an RMS torque error of 0.018 [N-m], or 0.14% of the maximum gravity load. The uncertainty in the gravity compensation is substantially larger than this. I originally believed that the gravity loading could be detected in the break-away data with sufficient sensitivity, by comparing data of the positive and negative rotation directions. Difference in the positive and negative rotation direction friction - still an enexplained phenomenon - upset this measurement however. For this work the gravity compensation was manually tuned to roughly 0.1 [N-m] accuracy by tuning gliding motions for minimum velocity variation: a tedious and imprecise process.

To conduct three joint motions a more complex inertial model must be used and gravity compensation is required. The model has three parts:

A. Model of rigid body dynamics,   9 Parameters ;
B. Gravity compensation,   4 Parameters ;
C. Friction Compensation   12 Parameters + 6 Tables .

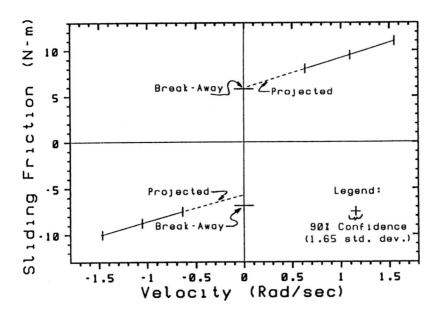

Negative Direction
Viscous 3.02 N-m/rad/sec
Kinetic 5.57 N-m

Positive Direction
Viscous 3.27 N-m/rad/sec
Kinetic 5.93 N-m

Figure 8.9.   Friction torque as a function of velocity, Joint 3.

The rigid body model used is the simplified model presented in [Armstrong, Khatib and Burdick 86]. The inertial, Coriolis and centrifugal forces are computed; this is feasible because the forces are pre-computed using the desired trajectory and applied open-loop. The lumped inertial parameters of the model depend upon the simple inertial parameters of each link, which are affected by sensors mounted on the arm. A difficult identification was anticipated, as the added sensing equipment will affect the nine lumped inertial parameters. The identification proved unnecessary; it was sufficient to measure the mass, and locate the center of mass of each of the added instruments. Using this information, the nine lumped inertial parameters were recomputed. Experience tuning the diagonal elements of the inertia matrix during motions of the individual joints suggests that the pre-computed parameters are accurate to ± 0.1 [N-m^2], or 2%.

The gravity parameters were tuned manually, as described above, and the friction parameters were those presented with figures 6.2, 8.8 and 8.9. Taken together they form a 25 parameter model which was used to produce the open-loop motions of figure 8.11. The cumulative position error at the

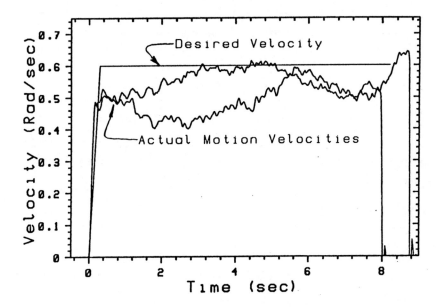

Figure 8.10   Velocity Profile Recorded During an Open-Loop Motion of Joint 3 with Position Dependent Correction Applied.

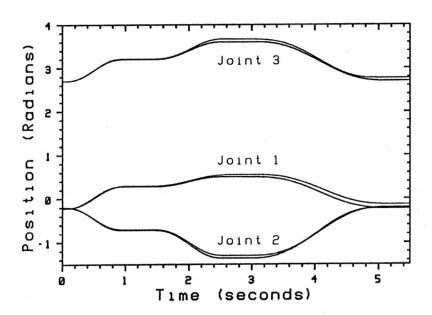

Figure 8.11   Position Profile Recorded During an Open-Loop Motion of Joints 1, 2 and 3.

end of the three joint, open-loop motion is less than 10%. Figure 6.1 is a photograph taken during the three joint open-loop motions. The pointers indicate the target points in space. During one three joint move a 50 gram weight as added to the end effector, creating an unmodeled disturbance. The impact was a ten fold increase in motion error.

8.3 Friction Compensated Force Control

To push the fidelity of force control to a level far below the magnitude of the static friction, an impulsive control system was implemented. Friction at very low velocities works against good control in three ways:

- When the mechanism is in static friction, the force across the friction junction is indeterminate;

- Because of the Dahl effect, break-away is not simple;

- Because of the Stribeck effect, motion at low velocities is unstable.

The manipulation of delicate objects in hard contact often occurs at low velocities. The low velocity motion arises because moderate changes in commanded force, or "force moves" in the parlance of force control, translate to tiny movements of a stiff system. In many practical situations the velocities will lie in the range of the Stribeck effect. For the mechanism studied, no feedback control is available which will stabilize the system in the regime of mixed lubrication, from 0.002 to 0.015 radians per second. This implies that any control that must cross this regime must do so unstably.

The motivation and design of the impulsive controller are heuristic. The impulsive controller operates by sending a series of short command sequences, called the *Impulse Torque*, which are tuned to achieve the desired change in contact force. The torque sequences are tuned empirically during a calibration process and loaded with the corresponding force step into a lookup table. The torque sequences used in these trials are presented in table 8.3. The impulse is combined with command feed forward and position-dependent friction compensation so that the total total applied control is give by:

$$Applied\,Torque\; =\; Impulse\,Torque\; +\; Desired\,Force\,Feed\,Forward$$
$$+\; Position\,Dependent\,Friction\,Compensation$$

The impulsive torque sequence is selected by measuring the force error, scaling the error by a gain and choosing the sequence corresponding to the largest force step smaller than the scaled force error. For example, with the gain set to 0.8, a desired contact force of 0.1 [N] and measured force of 0.05 [N] would give a force error of -0.05 [N] and a scaled force error of (desired force step) of -0.04 [N]. The system would choose the sequence 13-10, corresponding to a force step of 0.035 [N]. The system was calibrated to operate with a period of 5 mS, thus the two torques form a pulse lasting 10 mS. To these torques the desired contact force are added the feed forward, 0.1 [N] scaled by the radius, and the position-dependent friction compensation. The position-dependent friction compensation table was constructed as described above, with the spatial low pass filter set to 500 cycles per radian. Because the impulsive action excites the flexibilities of the mechanism, the force measurements are low pass filtered and a settling time is allowed to pass before the next control action is taken. For these trials the settling time was set to 120 mS, as indicated in table 8.3. A block diagram of the controller is shown in figure 8.12.

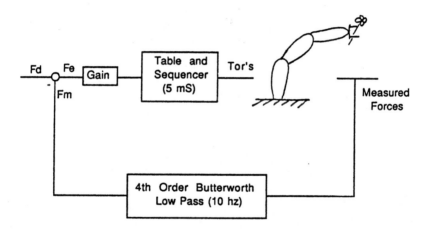

Figure 8.12   A Block Diagram of the Impulsive Controller.

The impulsive controller was used to insert a piece of wire wrap wire, #30 copper, 0.25 mm diameter, into a hole in a plate of glass, 0.75 mm diameter. The insertion posses a substantial control challenge because the buckling strength of the wire is only 0.2 Newtons, one 60th of the 12 Newton (radius scaled) static friction of the mechanism. The configuration of the apparatus is shown in figure 8.13. Joints 2 and 3 of the PUMA were position controlled by the standard industrial controller; they were used to move the wire along the 'V' slot between the glass slides. Joint 1 was aligned with the force control direction and was controlled by the impulsive controller.

Table 8.3   Impulse Table for the Impulsive Controller.

Force Step [Newtons]	Torques in Sequence	Settling Time [sec]	First Torque [N-m]	Second Torque [N-m]	Third Torque [N-m]
0.000	1	0.005	0		
0.020	1	0.12	13		
0.035	2	0.12	13	10	
0.055	2	0.12	13	13	
0.076	2	0.12	14	14	
0.096	3	0.12	15	15	2

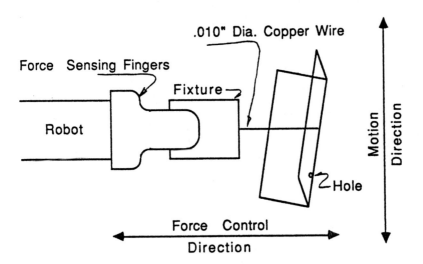

Figure 8.13   Experimental Setup for the Wire Wrap Insertion.

The measured force, control action and measured position during a successful insertion are shown in figures 8.14, 8.15 and 8.16. The control action plotted in figure 8.14 is the force step corresponding to the impulse selected by the controller. The selected force step has been offset to -0.2 [N-m] so that the it may be presented with the actual and commanded forces. With a new wire, the insertion was successful in roughly half of the trials. The surface being tracked was canted with respect to the direction of motion, so that travel in the force controlled direction was required to maintain contact, as indicated by the change in position between t = 0 and t = 5 seconds in figure 8.16. The wire was slanted into the direction of motion, and twice caught small features on the glass surface. These two points are

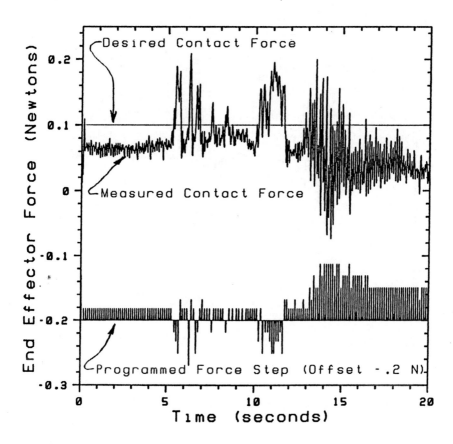

Figure 8.14   Measured Force During Wire Wrap Wire Insertion.

indicated by the rise of the force error to positive values, at these points the control system reverses its action. The force sensing fingers, wire wrap wire and contact surface a pictured in figure 8.17. This picture was taken at t = 11 seconds when the wire was experiencing its maximum deflection. The applied force at t = 11 is 0.18 Newtons, roughly the weight of 3 of these pages. At t = 12 seconds insertion occurs and the control system drives the wire through the hole.

The commanded force during this motion was 0.1 Newtons, one 120th of the static friction; the RMS contact force error is 0.038 Newtons, one 316th of the static friction. The RMS control action is 2.76 Newtons, 73 times greater than the RMS applied force error.

To account for the variation in friction with position, table lookup friction compensation was applied during impulsive control. For this application, a

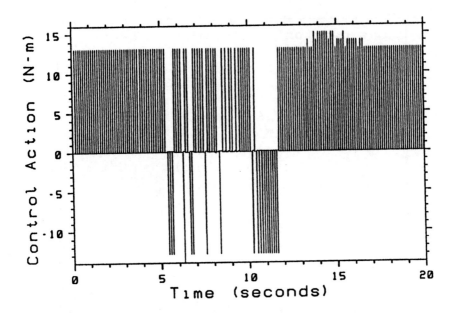

Figure 8.15   Control Action During Wire Wrap Wire Insertion.

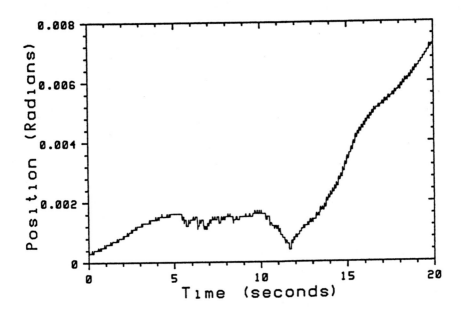

Figure 8.16   Measured Position During Wire Wrap Wire Insertion.

Figure 8.17   PUMA Arm Inserting Wire Wrap Wire Under Force Control.

table low pass filtered to 500 cycles per radian was used. The contribution of the table lookup compensator to the motion of figures 8.14, 8.15 and 8.16 is shown in figure 8.18. Compensation for the position-dependent friction is especially important for the smallest force steps, those corresponding to 0.020, 0.035 and 0.055 Newtons. A diminishment in these impulses of 1 [N-m] corresponds to a reduction in the output of a factor of 3; an increase of 1 [N-m] corresponds to an overshoot of a factor of 2. In figure 8.18, the regions of rapid reversal near six and eleven seconds correspond to reversals in control action evident in figures 8.15 and 8.16. A reversal in desired motion direction causes a switch from the positive direction table to the negative direction position-dependent friction compensation table.

   In this chapter three demonstrations of friction-compensated motion are presented. The demonstration of open-loop single joint motion shows that the friction model can be made quite accurate. The demonstration of three joint motion shows that the modeling techniques developed for joint one of the PUMA mechanism extend to joints two and three, and points the way toward maximum performance motion of the mechanism. It should be noted that during the motions presented, the inertial forces at no time exceed the

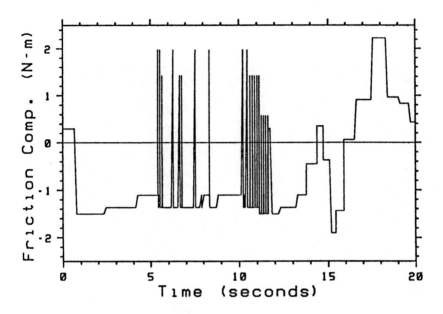

Figure 8.18   Position-Dependent Friction Compensation During the Force-Controlled Insertion.

static friction; that is to say that if the same motions had been undertaken with torques pre-computed using only an inertial model, no motion whatever would have occurred. A demonstration of force control is also presented, using a special 'impulsive' controller. The control law of this controller specifies very hard, quick actions to overcome friction. This control law is applicable only near zero velocity and suffers several limitations; but in its special application - high fidelity control of forces during low speed motions of a mechanism with substantial static, Dahl and Stribeck friction - it achieves a friction/force error ratio that is substantially better than the best previously demonstrated.

Chapter 9

Suggestions Toward Friction Modeling and Compensation

"Therefore always when you wish to know the quantity of the force that is required in order to drag the same weight over beds of different slope, you have to make the experiment and ascertain what amount of force is required to move the weight along a level road, that is to ascertain the nature of its friction."

Leonardo da Vinci (1452-1519),
The Notebooks, F II 106 r

Friction is a performance limiting factor in many mechanisms. By modeling friction these limits can be understood. This work points to a number of frictional properties which should be measured and suggests experimental procedures for obtaining the measurements. The properties of mechanism friction which should be measured are:

1. Repeatability of the friction forces;

2. Magnitude of the kinetic friction;

3. Magnitude of the static friction ;

4. Characteristic velocity of the Stribeck friction;

5. Time Lag in the change of sliding friction, indicated by τ_L;

6. Interaction of dwell time and the magnitude of static and Stribeck friction, indicated by γ;

7. Linearity of friction as function of velocity and the magnitude of the viscous friction;

8. Magnitude and character of position dependence in the static and kinetic friction.

9.1 Suggestions on Experimental Technique

With a force-actuated mechanism, it is straightforward to measure the repeatability of friction forces and the magnitude of kinetic and viscous friction. A stiff velocity servo should be implemented, as in section 6.2, and constant velocity motions conducted at a range of speeds; during these motions the average required torque should be measured. Several samples should be collected with each starting point and velocity in order to estimate the repeatability; such issues as warm up and independence from load should be examined. By reducing the commanded speed during quasi-constant velocity motion, the speed at which stick-slip motion begins can be observed. This velocity is not $\dot{x}_s$ of equation (6.14), but the experiment is straight forward and will provide an estimate useful in establishing the limits to performance.

Because the static friction can vary with position at very high spatial frequency, the magnitude of the static friction must be measured with care. The break-away experiment of chapter 5 will measure local static friction and can be implemented on any apparatus with force-commanded actuation and position sensing. The keys to successful measurement of static friction are sampling at adequate spatial frequency and binning according to position. If the sampling is not sufficiently dense, problems of undersampling and aliasing arise. If the data are examined according to the natural distribution rather than evenly-spaced bins, non-uniform friction will result in non-uniform sampling and bias the estimate of static friction. Position dependence is introduced into the friction by the non-uniformity of actuators and transmission elements. A spatially inhomogeneous transmission was the dominant source of position-dependent friction in the mechanism studied here. A mechanism that employed uniform actuation and transmission, such as pneumatic actuation through a four bar linkage or the cable driven manipulator of Salisbury and Townsend *et. al.* [88] might not exhibit significant spatial dependency. This work has shown that a dc servo motor and high ratio gear drive can exhibit a 30% change in static friction over a distance small in relation to the smallest feature of the actuator or drive train. Sampling with a spatial frequency 20 times higher than the finest gear pitch or the expected motor cogging is recommended.

Properties 5 and 6 are the time lag in sliding friction, τ_L and the dependence of static friction on dwell time, γ. These properties are more difficult than the others to measure directly. Direct measurement of the frictional lag requires measurement of instantaneous friction and thus acceleration sensing. Even with accelerometry, position-dependent friction and machinism flexibility may confound direct measurement in practical servo mechanisms. Direct measurement of γ is appears less sensitive to mechanism behaviors than measurement of τ_L, but instantaneous friction measurement is none-the-less required. The compliant motion experiments of sections 6.3 and 6.6 demonstrate one suitable procedure.

Identification may be a better approach to obtaining τ_L and γ. An experiment is suggested by figure 7.8. By mapping the regions of stick-slip and smooth motion using compliant motions, various compliances and a range of force rates, the stick-slip extinction boundary can be identified. At high velocities the stick-slip extinction is dominated by the diminishing dwell time, and γ can be determined via equation (7.36). At high stiffness and low velocity, stick-slip extinction is dominated by the frictional lag, and τ_L can be determined via equation (7.17), or approximated via equation (7.26). When the stiffness-velocity plane is sampled with sufficient density, the characteristic velocity of the Stribeck friction, $\dot{x}_s$, can be estimated by observing the desired velocity most difficult to stabilize, corresponding to the peak of figure 7.7. The sampling of such an experiment should be more dense than the sampling of figure 7.8. The 17 stiffness-velocity combinations of figure 7.8 are not sufficient to accurately identify $\dot{x}_s$; but could serve to place bounds on its possible range.

9.2 Suggestions on Control

For control, friction may be divided into two regimes: near zero velocity and away from zero velocity. Away from zero velocity the problem is a straightforward one: the friction should be modeled, and the model used in feedback design or for feed-forward compensation. The experience of this research shows friction in the moderate to high velocity regime (physically, the full fluid lubrication regime) to be well behaved. There is some impetus to neglect friction in this velocity regime because it is benign: it is never destabilizing. When friction is substantial relative to the inertial forces, it is at best faineant to neglect friction in control design. Mechanism damping should be understood and accurately considered in control design. For multi-degree-of-freedom mechanisms with coordinated control, such as robotic manipulators commanded by operational space or Salisbury stiffness control, unmodeled friction will alter the coupling between degrees-of-freedom and reduce the quality of coordination. For single degree-of-freedom mechanisms, the neglect of friction will result in an overly conservative design and an unnecessarily diminished performance.

When stick-slip is an issue, one should check the oil. We have seen that the occurrence and elimination of stick-slip is dominated by the temporal phenomena of friction. The literature shows that these phenomena can vary markedly with changes in lubricant selection. The central lesson of chapter 2 is that lubricant additives span a great spectrum and serve many purposes, among which friction modification may not have received the attention it deserved. In the field of lubrication engineering, friction modification has been dominated by wear reduction and issues of lubricant life and delivery. Indeed, one can find texts and surveys of the subject which neglect friction modification altogether. Thus, although 'checking the oil' is not normally the lot of the controls engineer, by insisting on a lubricant with an effective friction modifier, the controls engineer may make considerable progress toward achieving smooth motion. Constituting perhaps a percent of the total lubricant, this ounce of prevention can offset an expensive cure.

When friction modification has reached its limit and the treatment of stick-slip by feedback control is necessary, stiff and perhaps variable structure control are indicated. The analysis of chapter 7 indicates that the limit-cycle period relative to frictional lag (governed by system and control stiffness) and damping relative to excitation (governed by system and control damping) most directly extinguish stick-slip. The low frequency stiffness of integral control is found to be relatively ineffective in extinguishing stick-slip. Directly sensing friction, by sensing the torque or force output of the mechanism, can assist in achieving the necessary compensation.

The non-linear nature of low-velocity friction complicates control design, but in the final analysis it is the ordinary barriers to stiff control - sensor and actuator bandwidth limitations, unmodeled flexibility and computational delay - which set the limits on low speed motion. Issues of implementation will set the limits on control. And the friction model, with experimentally determined parameters, will determine the interaction of control and stick-slip. With knowledge of the system dynamics, limits on control and parameters of the friction model, the analysis of chapter 7 can be used to determine the lowest governable velocity and the smallest possible motion, and thus the lower performance bounds of the machine.

9.3 Conclusion

This work has shown that friction may be substantial and quite repeatable, and that it can be effectively predicted and compensated during motion. In the particular mechanism studied, a dc servo motor driven rotary joint with grease lubricated gears and ball bearings, it was found that:

- Motion friction depends upon position and is about 99% repeatable;
- Break-away friction, when measured in a controlled way, is about 97% repeatable;
- Above a minimum velocity, the friction is proportional to velocity;
- The Stribeck effect plays a destabilizing role below a minimum velocity;
- The Dahl Effect plays an important role in the transition from rest to motion;
- A change in friction lags behind a change in velocity, and the frictional lag dominates the interaction between stiffness and stick-slip;
- Static friction rises with time at zero velocity, and the rising static friction dominates the interaction between velocity and stick-slip;
- The friction is different in the positive and negative rotation directions.

Demonstrations of open-loop control and high-fidelity force control show that friction modeling can be quite accurate.

Though the experimental results of this work apply only to a particular mechanism, the methodology and general findings are more widely applicable. The availability of a tribologically supported model of the Stribeck effect, frictional lag, and rising static friction has permitted the development of analytic and experimental techniques to study and measure these effects and their impact on control. Accurate friction modeling will facilitate effective off-line tuning of control systems, implementation of optimal control strategies in friction affected mechanisms, and accurate simulation of motion. An understanding of the relationship between lubricant properties and low velocity dynamics will enable appropriate lubricant selection, control design to minimize stick-slip and, in some cases, control that achieves smooth motion from zero to the maximum velocity. These are objectives of control which may be achieved by accurate friction modeling: by the study of the machine *to ascertain the nature of its friction.*

Bibliography

Amontons, G. 1699, "On the Resistance Originating in Machines," *Proceedings of the French Royal Academy of Sciences*, p. 206-22.

Armstrong-Helouvry, B. 1990 (May), "Stick-Slip Arising from Stribeck Friction," *Proc. 1990 Inter. Conf. on Robotics and Automation*, Cincinnati: IEEE, 1377-82.

Armstrong, B. 1989 (June), "Control of Machines with Non-Linear Low-Velocity Friction: A Dimensional Analysis," *Proceedings of the First International Symposium on Experimental Robotics*, Montreal, Quebec, p. 180-95.

Armstrong, B. 1989a, "On Finding Exciting Trajectories for Identification Experiments Involving Systems with Nonlinear Dynamics," *Int. J. of Robotics Research*, 8(6):28-48.

Armstrong, B. 1988, "Dynamics for Robot Control: Friction Modeling and Ensuring Excitation During Parameter Identification," *PhD Thesis*, Dept. of Electrical Engineering, Stanford University, May 1988; Stanford Computer Science Memo STAN-CS-88-1205.

Armstrong, B. 1988a (May), "Friction: Experimental Determination, Modeling and Compensation," *Proc. 1988 Inter. Conf. on Robotics and Automation*, Philadelphia: IEEE, p. 1422-7.

Armstrong, B., Khatib, O. and Burdick, J. 1986 (April 7-10), "The Explicit Dynamic Model and Inertial Parameters of the PUMA 560 Arm," *Proc. 1986 Inter. Conf. of Robotics and Automation*, San Fransisco: IEEE, 510-518

Asada, H. and Youcef-Toumi, K., 1984, "Analysis and Design of a Direct-Drive Arm with a Five-Bar Parallel Drive Mechanism," *ASME Journal of Dynamic Systems, Measurment and Control*, vol. 106, p. 225-230.

Atherton, D.P. 1975, *Nonlinear Control Engineering*, London: Van Nostrand Reinhold Co. Ltd.

Bell, R. and Burdekin, M. 1969, "A Study of the Stick-Slip Motion of Machine Tool Feed Drives," *Proc. of the Instn. of Mechanical Engineers*, vol. 184, pt. 1, no. 29, p. 543-60.

Bell, R. and Burdekin, M. 1966, "Dynamic Behavior of Plain Slideways," *Proc. Instn. of Mechanical Engineers,* vol. 181, pt. 1, no. 8, p. 169-83.

Bennett, S. 1979, *A History of Control Engineering,* Sterenage: Peter Peregrinos LTD.

Biel, C. 1920, "Die Reibung in Glietlagern bie Zusatz von Voltool zu Mineralol und bie Veranderung der Unlaufzahl und der Temperatur," *Zeitschrift des Vereines Seutscher Ingenieure,* 64(1920):449-83.

Bo, L.C. and Pavelescu, D. 1982, "The Friction-Speed Relation and Its Influence on the Critical Velocity of the Stick-Slip Motion," *Wear,* 82(3):277-89.

Bohacek, P.K. and Tuteur, F.B. 1961 (May), "Stability of Servomechanisms with Friction and Stiction in the Output Element," *IRE Transactions on Automatic Control,* AC-6(61):222-27.

Booser, E.R. (ed.) 1984, *CRC Handbook of Lubrication,* Boca Raton: CRC Press.

Bowden, F.P. and Tabor, D. 1973, *Friction - An Introduction to Tribology,* New York: Anchor Press/Doubleday; Reprinted 1982, Malabar: Krieger Publishering Co.

Bowden, F.P. and Tabor, D. 1956, *Friction and Lubrication,* New York: John Wiley and Sons Inc.

Bowden, F.P. 1950, *BBC Broadcast.*

Bowden, F.P. and Leben, L. 1939, "The nature of Sliding and the Analysis of Friction," *Proc. of the Royal Society, Series A,* vol. 169, p. 371-91.

Bowden, F.P. and Tabor, D. 1939, "The Area of Contact Between Stationary and Between Moving Surfaces," *Proc. of the Royal Society, Series A,* vol. 169, p. 391-413.

Brandenburg, G., Hertle, H., and Zeiselmair, K. 1987 (July), "Dynamic Influence and Partial Compensation of Coulomb Friction in a Position- and Speed-Controlled Elastic Two-Mass System," *Proc. 10th World Congress on Automatic Control,* IFAC: Munchen, vol. 3, p. 91-99.

Brandenburg, G., and Schafer, U. 1988 (September), "Influence and Partial Compensation of Simultaneously Acting Backlash and Coulomb Friction in a Position- and Speed-Controlled Elastic Two-Mass System," *Proc. 2nd Int. Conf. on Electrical Drives,* ICED: Poiana Brasov.

Brandenburg, G., and Schafer, U. 1990 (January), "Model Reference Position Control of an Elastic Two-Mass System with Backlash and Coulomb Friction Using Different Types of Observers," *Power Electronics and Motion Control,* PEMC: Budapest.

Brockley, C.A., Davis, H.R. 1968, "The Time-Dependence of Static Friction," *J. of Lubrication Technology,* 90(1):35-41.

Brockley, C.A., Cameron, R., and Potter, A.F. 1967, "Friction-Induced Vibration," *J. of Lubrication Technology,* 89(2):101-8.

Burdekin, M., Back, N. and Cowley, A. 1978, "Experimental Study of Normal and Shear Characteristics of Machined Surfaces in Contact," *J. of Mechanical Engineering Science,* 20(3):129-32.

Cameron, A. 1984 (September), "On a Unified Theory of Boundary Lubrication," *Proc. of 11th Leeds-Lyon Symp. on Tribology, Leeds,* London: Butterworths.

Cannon, R. and Schmitz, E. 1984, "Initial Experiments on the End-Point Control of a Flexible One-Link Robot," *Int. J. of Robotics Research,* 3(3)62:75.

Canudas de Wit, C. 1988, *Adaptive Control for Partially Known Systems - Theory and Applications,* Amsterdam: Elsevier.

Canudas de Wit, C., Astrom, K.J., and Braun, K. 1987, "Adaptive Friction Compensation in DC Motor Drives," *IEEE J. of Robotics and Automation,* RA-3(6).

Canudas de Wit,C., Noel,P., Aubin,A., Brogliato,B. and Drevet,P. 1989 (May), "Adaptive Friction Compensation in Robot Manipulators: Low-Velocities," *Proc. Inter. Conf. on Robotics and Automation,* Scottsdale: IEEE, 1352-7.

Canudas de Wit, C. and Seront, V. 1990 (May), "Robust Adaptive Friction Compensation," *Proc. Inter. Conf. on Robotics and Automation,* Cincinnati: IEEE, 1383-9.

Cheng, J.H. and Kikuchi, N. 1985, "An Incremental Constitutive Relation of Unilateral Contact Friction for Large Deformation Analysis," *J. of Applied Mechanics,* 52(3):639-48.

Cincinnati Milacron 1986, *Revised Stick-Slip Test Procedure.*

Coulomb, C.A. 1785, "Théorie des machines simples, en ayant égard au frottement de leurs parties, et a la roideur dews cordages," *Mém. Math Phys.,* Paris, vol. x, p. 161-342.

Craig, J.J. 1987, *Adaptive Control of Mechanical Manipulators,* Menlo Park: Addison-Wesley Publishing Company.

Craig, J.J. 1986, "Adaptive Control of Mechanical Manipulators," *PhD Thesis,* Electrical Engineering Dept., Stanford University.

Czichos, H. 1978, *Tribology,* Amsterdam: Elsevier Scientific Pup. Co.

Da Vinci, L. (1519), *The Notebooks,* Ed. Jean Paul Richter, New York: Dover Pub. Inc. (1970).

Dagalakis, N.G. and Myers, D.R. 1985, "Adjustment of Robot Joint Gear Backlash Using the Robot Joint Test Excitation Technique," *Int. J. of Robotics Research,* 4(2):65-79.

Dahl, P.R. 1977, "Measurement of Solid Friction Parameters of Ball Bearings," *Proc. of 6th Annual Symp. on Incremental Motion, Control Systems and Devices,* U. of Illinois.

Dalh, P.R. 1976, "Solid Friction Damping of Mechanical Vibrations," *AIAA J.,* 14(12):1675-82.

Dahl, P.R. 1968, "A Solid Friction Model, " TOR-158(3107-18), The Aerospace Corporation, El Segundo, California.

Derjaguin, B.V., Push, V.E. and Tolstoi, D.M. 1957 (October), "A Theory of Stick-Slip Sliding of Solids," *Proceedings of the Conference on Lubrication and Wear,* London: Instn. Mech. Engs, p. 257-68.

Dokos, S.J. 1946 (June), "Sliding Friction Under Extreme Pressures- 1," *J. of Applied Mechanics,* vol. 13, series A, p. 148-56.

Dowson, D. 1979, *History of Tribology,* London: Longman Ltd.

Dowson, D. and Higginson, G.R. 1966, *Elasto-Hydrodynamic Lubrication - The Fundamentals of Roller and Gear Lubrication,* Oxford: Pergamon Press.

Dowson, D., Taylor, C.M., Godet, M. and Berthe, D. (eds.) 1980, "Friction and Traction," *Proceedings of the 7th Leeds-Lyon Symp. on Tribology,* Guildford: Westbury House.

Dudley, B.R. and Swift, H.W. 1949, "Frictional Relaxation Oscillations," *Philosophical Magazine,* Vol. 40, Series 7, p. 849-61.

Dupont, P.E. 1990 (May), "Friction Modeling in Dynamic Robot Simulation," *Proc. 1990 Inter. Conf. on Robotics and Automation,* Cincinnati: IEEE, 1370-7.

Dupont, P.E. 1991 (April), "Avoiding Stick-Slip in Position and Force Control Through Feedback," *Submitted to the 1991 Inter. Conf. on Robotics and Automation,* Sacramento: IEEE.

Estler, R. B. 1980 (October), "What's New in Preventive Maintenance for Gears and Bearings," *Conf. of the American Society of Lubrication Engineers,* Baltimore: ASLE.

Facchiano, D.L. and Vinci, J.N. 1984, "EP Industrial Gear Oils - A Look at Additive Functions and a Comparison of Sulfur Phosphorus and Leaded Gear Oils," *Lubrication Engineering,* 40(10):598-604.

Fearing, R. 1987, Private Communication.

Friedland, B. 1990, "Adaptive Friction Compensation," *Submitted to the 1990 Conf. on Decision and Control,* IEEE.

Fuller, D.D. 1984, *Theory and Practice of Lubrication for Engineers,* New York: John Wiley and Sons.

Gassenfeit, E.H. and Soom, A. 1988, "Friction Coefficients Measured at Lubricated Planar Contacts During Start-Up," *J. of Tribology,* 110(3):533-538.

Gilbart, J.W. and Winston, G.C. 1974, "Adaptive Compensation for an Optical Tracking Telescope," *Automatica,* 10(2):125-31.

Gogoussis, A. and Donath, M. 1987 (March 31 - April 3), "Coulomb Friction and Drive Effects in Robot Mechanisms," *Proc. of the 1987 Inter. Conf. on Robotics and Automation,* Raleigh: IEEE, 828-36.

Haessig, D.A. and B. Friedland (May 1990), "On the Modeling and Simulation of Friction," *Proc. of the 1990 American Control Conference,* ASME: San Diego; To Appear in the *J. of Dynamic Systems, Measurement and Control.*

Halling, J. 1975, *Principles of Tribology,* London: Macmillan Publishing Ltd.

Hamrock, B.J. 86, "Lubrication of Machine Elements," In the *Mechanical Engineers' Handbook,* Kutz, M., Ed., New York: John Wiley and Sons Inc.

Hersey, M.D. 1966, *Theory and Research in Lubrication,* New York: John Wiley and Sons, Inc.

Hersey, M.D. 1914, "The Laws of Lubrication of Horizontal Journal Bearings," *J. Wash. Acad. Sci.,* vol. 4, p. 542-52.

Hertz, H. 1881, "On the Contact of Elastic Solids," *J. Reine und Anges. Math.,* vol. 92, p. 156-71.

Hess, D.P. and Soom, A. 1990, "Friction at a Lubricated Line Contact Operating at Oscillating SLiding Velocities," *J. of Tribology,* 112(1):147-52.

Johnson, E.C. 1952, "Sinusoidal Analysis of Feedback Control Systems Containing Non-Linear Elements," *Trans. Amer. Inst. of Elect. Engineers.* vol. 71, pt. 2, no. 1, p. 169-81.

Karlen, J.P., Thompson, J.M., Void, H.I., Farrell, J.D., and Eismann, P.H. 1990, "A Dual-Arm Dexterous Manipulator System with Anthropomorphic Kinematics," *Proc. 1990 Inter. Conf. on Robotics and Automation,* Cincinnati: IEEE, 368-73.

Kato, S., Sato, N. and Matsubay, T. 1972 , "Some Considerations of Characteristics of Static Friction of Machine Tool Slideway," *J. of Lubrication Technology,* 94(3):234-47.

Kelly, R. 1987, "A Linear-State Feedback Plus Adaptive Feed-Forward Control for DC Servomotors," *IEEE Trans. on Industrial Electronics,* vol. IE-34, no. 2 p. 153-7.

Khatib, O. and Burdick, J. 1986 (April 7-10), "Motion and Force Control of Robot Manipulators," *Proc. 1986 Inter. Conf. of Robotics and Automation,* San Fransisco: IEEE, 1381-6.

Khitrik, V.E. and Shmakov, V.A. 1987, "Static and Dynamic Characteristics of Friction Pairs," *Soviet J. of Friction and Wear,* 8(5):112-5.

Ko, P.L. and Brockley, C.A. 1970, "The Measurement of Friction and Friction-Induced Vibration," *J. of Lubrication Technology,* 92(4):543-9.

Kubo, T., Anwar, G. and Tomizuka, M. 1986 (April 7-10), "Application of Nonlinear Friction Compensation to Robot Arm Control," *Proc. 1986 Inter. Conf. of Robotics and Automation,* San Fransisco: IEEE, 722-7.

Luh, J.Y.S., Fisher, W.D. and Paul, R.P.C. 1983, "Joint Torque Control by a Direct Feedback for Industrial Robots," *IEEE Trans. on Automatic Control,* AC-28(2):153-61.

Lubrizol, Inc. 1988, Product Data Sheet, Lubrizol 5346.

Ludema, K.C. 1988, "Engineering Progress and Cultural Problems in Tribology," *Lubrication Engineering,* 44(6):500-9.

Maples, J. 1985(June), "Force Control of Robotic Manipulators with Structural Flexibility," *PhD Thesis,* Electrical Engineering Dept., Stanford University.

Merchant, M.E. 1946, "Characteristics of Typical Polar and Non-Polar Lubricant Additives Under Stick-Slip Conditions," *Lubrication Engineering,* 2(2):56-61.

Millman, G. 1990, Cincinnati Millacron, Private Communication.

Mobil Oil Corporation 1978, "Way Lubrication - Machine Tools," Technical Bulletin 8-93-001.

Mobil Oil Corporation 1971, "Gears and Their Lubrication," Technical Bulletin 1-92-003.

Morin, A. J. 1833, "New Friction Experiments carried out at Metz in 1831-1833," *Proceedings of the French Royal Academy of Sciences,* vol. 4, p. 1-128.

Mukerjee, A. and Ballard, D.H. 1985 (March), "Self-calibration in Robot Manipulators," *Proc. 1985 Inter. Conf. on Robotics and Automation,* St. Louis: IEEE, p. 1050-7.

Newton, I. (1687), *Philosophiae Naturales Principia Mathematica,* S. Pepys, Reg. Soc. Praeses, 5 Julii, 1686.

O'Connor, J.J., Boyd, J. and Avallone, E. A. (Eds.) 1968, *Standard Handbook on Lubrication Engineering,* New York: McGraw Hill.

Oldenberger, R. and Boyer, R.C. 1962, "Effects of Extra Sinusoidal Inputs to Nonlinear Systems," *Trans. A.S.M.E., J. of Basic Engineering* D-84():559-70.

Ostachowicz, M.W. 1987 (September), "The Effect of Coulomb Friction in Periodic Motion of Industrial Robots," *Proc. Congress on the Theory of Machines and Mechanisms,* Sevilla: IFToM.

Pan, P. and Hamrock, B.J. 1989, "Simple Formulas for Performance Parameters Used in Elastohydrodynamically Lubricated Line Contacts," *J. of Tribology,* 111():246-51.

Papay, A.G. 1988, "Industrial Gear Oils - State of the Art," *Lubrication Engineering,* 44(3):218-29.

Papay, A.G. 1974, "EP Gear Oils Today and Tomorrow," *Lubrication Engineering,* 30(9):445-54.

Papay, A.G. and Dinsmore, D.W. 1976, "Advances in Gear Additive Technology," *Lubrication Engineering,* 32(5):229-34.

Pavelescu, D. and Tudor, A. 1987, "The Sliding Friction Coefficient - Its Evolution and Usefulness," *Wear,* 120(3):321-36.

Pfeffer, L., Khatib, O. and Hake, J. 1986 (June), "Joint Torque Sensory Feedback in the Control of a PUMA Manipulator," *Proc. 1986 American Control Conference,* ASME: Seattle.

Pfeffer, L., Khatib, O. and Hake, J. 1989, "Joint Torque Sensory Feedback in the Control of a PUMA Manipulator," *IEEE Trans. on Robotics and Automation,* 5(4):418-25.

Rabinowicz, E. 1978, "Friction, Especially Low Friction," *International Conference on Fundimentals of Tribology,* (1978: MIT) edited by N.P. Suh and N. Saka, Cambridge: MIT Press, p. 351-65.

Rabinowicz, E. 1965, *Friction and Wear of Materials* New York: John Wiley and Sons.

Rabinowicz, E. 1958, "The Intrinsic Variables affecting the Stick-Slip Process," *Proc. Physical Society of London,* 71():668-75.

Rabinowicz, E. 1956, "Autocorrelation Analysis of the Sliding Process," *J. of Applied Physics,* 27(2):131-5.

Rabinowicz, E. 1956a (May), "Stick and Slip," *Scientific American,* 194(5):109-18.

Rabinowicz, E. 1951, "The Nature of the Static and Kinetic Coefficients of Friction," *J. of Applied Physics,* 22(11):1373-9.

Rabinowicz, E. and Tabor, D. 1951a, "Metallic transfer between sliding metals: an autoradiographic study," *Proc. of the Royal Soc. of London,* A, vol. 208, 455:75.

Rabinowicz, E. Rightmire, B.G., Tedholm, C.E. and Willians, R.E. 1955, "The Statistical Nature of Friction," *Trans. of the ASME,* 22():981-4.

Rayiko, M.V. and Dmytrychenko, N.F. 1988, "Some Aspects of Boundary Lubrication in the Local Contact of Friction Surfaces," *Wear,* 126(1):69-78.

Reynolds, O. 1886, "On the Theory of Lubrication and its application to Mr. Beauchamp Tower's experiments, including an experimental determination of the viscosity of olive oil," *Phil. Trans. Royal Soc.,* vol. 177, p. 157-234.

Rice, J.R. and Ruine, A.L. 1983, "Stability of Steady Frictional Slipping," *J. of Applied Mechanics,* 50():343-9.

Roberts, R.K. 1984, "The Compliance of End Effector Force Sensors for Robot Manipulator Control," *PhD Thesis,* Electrical Engineering Dept., Purdue University.

Rooney, G.T. and Deravi, P. 1982, "Coulomb Friction in Mechanism Sliding Joints," *Mechanism and Machine Theory,* 17(3):207-11.

Sadeghi, F. and Sui, P.C. 1989 (January), "Compressible Elastohydrodynamic Lubrication of Rough Surfaces," *J. of Tribology,* 111():56-62.

Sampson, J.B., Morgan, F., Reed, D.W. and Muskat, M. 1943, "Friction Behavior During the Slip Portion of the Stick-Slip Process," *J. Applied Physics,* 14(12):689-700.

Satyendra, K.N. 1956 (September), "Describing Functions Representing the Effect of Inertia, Backlash and Coulomb Friction on the Stability of an Automatic Control System," *AIEE Transactions,* vol. 75, pt. II, p. 243-9.

Salisbury, J.K., Townsend, W.T., Eberman, B.S., DiPietro, D.M. 1988 (April), "Preliminary Design of a Whole-Arm Manipulation System (WAMS)," *Proc. of the 1988 Int. Conf. on Robotics and Automation,* Philadelphia, Pa: IEEE, 254-60.

Shen, C.N. 1962, "Synthesis of High Order Nonlinear Control Systems with Ramp Input", *IRE Trans. on Automatic Control,* AC-7(2):22-37.

Shen, C.N. and Wang, H. 1964, "Nonlinear Compensation of a Second- and Third-Order System with Dry Friction," *IEEE Trans. on Applications and Industry,* 83(71):128-36.

Silverberg, M.Y. 1957, "A Note on the Describing Function of an Element with Coulomb, Static and Viscous Friction," *AIEE Trans.,* vol. 75, pt. II, p. 423-5.

Singh, S.K. 1990, "Modified PID Control with Stiction: Periodic Orbits, Bifurcation and Chaos," Report CAR-90-04, Dept. of Mechanical Engineering, Dartmouth University.

Sommerfeld, A. 1904, "Zur Hydrodynamischen Theorie der Schmiermittehreibung," *Zeitschrift Fur Mathematic und Physik,* 50(1904):97-155.

Sroda, P. 1988, "Analysis of the Shape of the Contact Geometry During Meshing of Involute Gears," *Wear,* 121(2):183-196.

Stribeck, R. 1902, "Die Wesentlichen Eigenschaften der Gleit- und Rollenlager - The Key Qualities of Sliding and Roller Bearings," *Zeitschrift des Vereines Seutscher Ingenieure,* 46(38):1342-48, 46(39):1432-37.

Thomas, S. 1930, "Vibrations Damped by Solid Friction," *Philosophical Magazine,* 7(9):329.

Threlfal, D.C. 1978, "The Inclusion of Coulomb Friction in Mechanisms Programs with Particular Reference to DRAM," *Mechanism and Machine Theory,* 13(4):475-83.

Tolstoi, D.M. 1967, "Significance of the Normal Degree of Freedom and Natural Normal Vibrations in Contact Friction," *Wear,* 10(3):199-213.

Tou, J. 1953, "Coulomb and Static Friction in Servo-Mechansisms," *PhD Thesis,* Electrical Engineering Dept., Yale University.

Tou, J. and Schultheiss, P.M. 1953, "Static and Sliding Friction in Feedback Systems," *J. of Applied Physics,* 24(9):1210-7.

Townsend, W.T. 1988, "The Effect of Transmission Design on the Performance of Force-Controlled Manipulators," *PhD Thesis,* Mechanical Engineering Dept., Massachusetts Institute of Technology.

Townsend, W. T. and Salisbury, J. K. 1987, "The Effect of Coulomb Friction and Sticktion on Force Control," *Proc. 1987 Inter. Conf. on Robotics and Automation,* Raleigh: IEEE, 883-9.

Truxal, J.G. 1955, *Automatic Feedback Control System Synthesis,* New York: McGraw-Hill Co., Inc.

Tustin, A. 1947, "The effects of Backlash and of Speed-Dependent Friction on the Stability of Closed-Cycle Control Systems," *IEE Journal,* vol. 94, part 2A, p. 143-51.

Villanueva-Leal, A. and Hinduja, S. 1984, "Modeling the Characteristics of Interface Surfaces by the Finite Element Method," *Proc. Instn. Mechanical Engineers,* 198C(4):9-23.

Vinogradov, G.V., Korepova, I.V. and Podolsky Y.Y. 1967, "Steel-to-Steel Friction Over a Very Wide Range of Sliding Speeds," *Wear,* 10(5):338-52.

Vischer, D. and Khatib, O. 1990, "Design and Development of Torque-Controlled Joints," *Experimental Robotics I,* eds. Hayward, V. and Khatib, O., Heidelberg: Springer-Verlag, p.271-86

Vischer, D. and Khatib, O. 1990a, "Performance Evaluation for Force/Troque Feedback Control Methodologies," *Proc. Romansy '90,* Cracow, Poland.

Walrath, C.D. 1984, "Adaptive Bearing Friction Compensation Based on Recent Knowledge of Dynamic Friction," *Automatica,* 20(6):717-27.

Wang, D. and Vidyasagar, M. 1987 (March 31-April 3), "Control of a Flexibile Beam for Optimum Step Response," *Proc. 1987 Inter. Conf. of Robotics and Automation,* Raliegh: IEEE, p. 1567-72.

Weaver, W. 1959, "Dither," *Science,* 130(3371):301.

Wellauer, E.J. and Holloway, G.A. 1976, "Application of EHD Oil Film Theory to Industrial Gear Drives," *J. of Engineering for Industry,* 98B(1):626-34.

Wills, G.J. 1980, *Lubrication Fundamentals,* New york: Marcel Dekker, Inc.

Wilson, A.R. 1979, "The Relative Thickness of Grease and Oil FIlms in Rolling Bearings," *Proc. Instn. Mechanical Engineers,* 193(17):185-92.

Wolf, G.J. 1965, "Stick-Slip and Machine Tools," *Lubrication Engineering,* 21(7):273-5.

Wu, C.H. and Paul, R.P. 1980 (December), "Manipulator Compliance Based on Joint Torque Control," *19th Conference on Decision and Control,* Albuquerque: IEEE, p. 88-94.

Xiaolan, A. and Haiqing, Y. 1987, "A Full Numerical Solution for General Transient Elastohydrodynamic Line Contacts and its Application," *Wear,* 121(2)143-159.

Zhu, D. and Cheng, H.S. 1988 (January), "Effect of Surface Roughness on Point Contact EHL," *J. of Tribology,* 110():32-7.

Appendix A

Small Studies

In this appendix the results of a number of small studies are reported. These small studies investigate friction properties that are not considered in the main thrust of the research but may none-the-less be of interest to some readers. The studies reported here are preliminary examinations made quickly with apparatus that was available.

A.1 Friction as a Function of Motor Angle

The break-away data of figure 5.2 show a position dependence in the static friction. Whether the observed structure in the static friction is connected with motor position can be tested by grouping the break-away data according to motor angle, as shown in figure A.1. To make figure A.1 the break-away data from a full rotation of joint 1, 52 motor revolutions, have been regrouped according to the motor angle. That is to say that the value plotted at each point in figure A.1 is the average of data collected at 52 different arm positions, all corresponding to one motor position. If the break-away friction were uncorrelated with motor angle, any structure in figure A.1 would be coincidental.

Figure A.1 shows twelve peaks in the friction per rotation of the motor. The pinion gear on the motor shaft is twelve pitch, suggesting that the periodic friction in figure A.1 occurs at one cycle per gear tooth.

The drive train of joint 1 of the PUMA 560 robot consists of two intermediate gears acting in parallel, the motor pinion gear and a large bull gear, as shown in figure A.2. If the friction plotted in figure A.1 is taken to be the friction in the motor and the motor/intermediate gear interface, it can be subtracted from the total friction to yield the residual friction. This was done and the residual friction was grouped according to intermediate gear rotation angle. The result, friction as a function of intermediate gear angle, is shown in figure A.3.

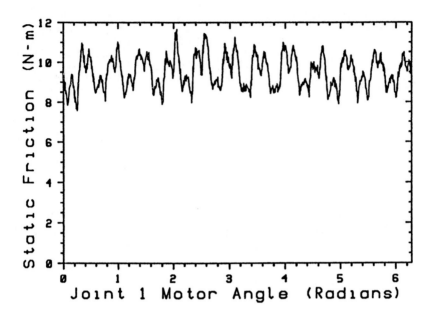

Figure A.1   Break-Away Torque as a Function of Motor Rotation Angle, merged data from 52 Motor Revolutions.

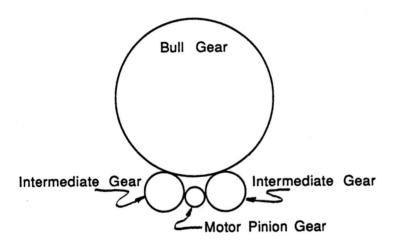

Figure A.2   Schematic Illustration of the arrangement of gears in joint 1 of the PUMA 560 robot.

The friction signal of figure A.3 shows a component at one cycle per intermediate gear revolution and another at 48 cycles per gear revolution. The one cycle per revolution component is due to eccentricity of the intermediate

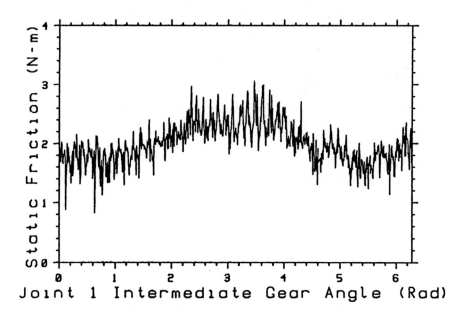

Figure A.3 Break-Away Torque as a Function of Intermediate Gear Rotation Angle, merged data from 13 Gear Revolutions.

gear. The intermediate gear is 48 pitch at the motor pinion/intermediate gear interface, accounting for the 48 cycle per revolution signal. The mean value of static friction in figure A.3 has been arbitrarily assigned to 2 Newton-meters: the experiment can not distinguish the DC value of static friction associated with each rubbing interface.

The 20% variation in friction (motion friction as well as static, see figure 5.1) occurring with the passage of each motor tooth is a very substantial disturbance to control. Consider that in a standard controller the integral control term will attempt to track the varying friction of figure A.1. The friction disturbance spans the spatial frequency spectrum from one cycle per arm revolution to hundreds of cycles per radian. This disturbance is a strong impetus toward homogeneous drive mechanisms, such as that proposed and examined in [Townsend 88].

A.2 Joint 2 Motor Alone and Joint 2 Link Alone

During a maintenance operation the motor of joint 2 was detached from link 2. This opportunity was used to measure the friction in motor 2 alone and in link 2 alone. The break-away experiment was used to measure the static motor friction, as described in chapter 5.1. With the rotational accelerometer attached, link 2 was lifted and allowed to swing under the influence of gravity. The velocity and position were estimated by integrating the acceleration signal and the friction parameters were estimated as described in chapter 6.1.

The break-away friction of the motor alone is compared to the break-away friction of the motor and joint in figure A.4. The mean static friction of the motor alone is 8.0 N-m (reflected to the joint) compared with a mean static friction of 12.7 N-m for the motor with drive train and link. The dominant spatial frequency of the motor with link curve in figure A.4 is 65 cycles per revolution of the motor. This frequency does not correspond to any known drive train feature.

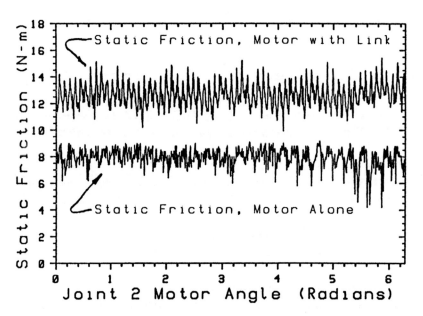

Figure A.4   Static Friction as a Function of Motor 2 Angle. Static Friction Measured with Link 2 Attached and Motor 2 Alone.

While link 2 was swinging freely, the link bearings and the intermediate gear were turning. The acceleration profile recorded during the swinging motion is shown in figure A.5; note the jumps in acceleration that occur when the velocity reverses. The kinetic and viscous friction parameters of the link alone are presented in figure A.6. The link bearings and bull gear/intermediate gear interface contribute 5% of the total kinetic and viscous friction.

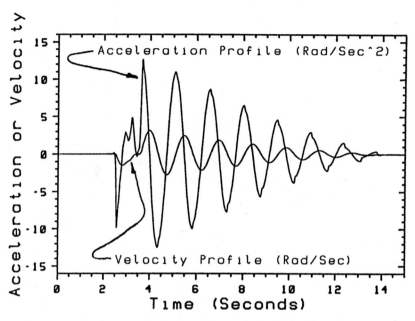

Figure A.5   Acceleration and Velocity Profile of Link 2 Swinging under the Influence of Gravity.

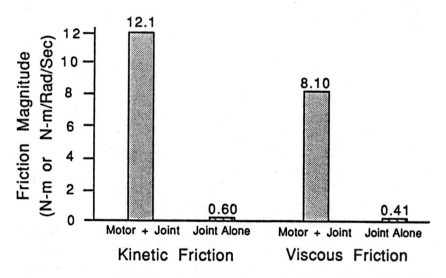

Figure A.6   Comparison of the Kinetic and Viscous Friction Parameters for the Assembled Joint 2 and for Link 2 without the Motor.

A.3 Trials with Dither

Dither is a high frequency signal added to the control signal; it is used to reduce the effect of static friction. Dither is commonly used in linear control of hydraulic actuators, where static friction can be very substantial. Dither was applied during force control in compliant contact and provided roughly a factor of three improvement in the fidelity of force control.

The apparatus used to test dither incorporated force feedback with a single pole lag compensator, and derivative feedback derived from the integrated acceleration signal. The force error gain was 8 Newton-meters per Newton and the derivative gain was 30 Newton meters per radian/second. The controller sampled 200 times per second. A stiff spring was used to give an environmental stiffness of 2,800 N-m per rad. The apparatus was configured as shown in figure 6.4.

In each trial a triangle wave was used as desired force command. Initial Contact occurred under active force control and was stable. The commanded and actual force from a trail with linear control and 4 Newton-meters of dither is shown in figure A.7. As seen in the figure, the actual force follows the desired force with an offset that is dependent on the derivative of desired force . The force error times the proportional gain is roughly the level of static friction, which averages 9.4 Newton-meters. Note that the use of integral control is limited in situations with non-linear fiction because of the tendency to induce limit cycling.

Two Newton-meter dither was applied at a range of frequencies, the RMS force error at each frequency is shown in figure A.8. With a controller sample rate of 200 Hz, the maximum - and most effective - dither frequency is 100 Hz. Dither was next applied at 100 Hz and a range of amplitudes. The efficacy of dither as a function of amplitude is presented in figure A.9.

Figure A.9 shows dither to reduce RMS force error, even at amplitudes substantially greater than the static friction of the mechanism. The motor currents applied to achieve force control with 12 N-m of dither at 100 Hz are shown in figure A.10. The robot sang audibly with dither of 12 N-m or more. The fidelity of force control with dither and the non-linear impulsive control of section 8.3 are compared in figure A.11. To motions are evaluated in the comparison: one a force command peaking at 8 Newtons, as shown in figure A.7; the other an identical force profile, scaled down by a order of magnitude. At the higher force level dither and the impulsive control give comparable performance. At the lower force level the impulsive controller yields the lower RMS force error.

Neither dither nor the impulsive controller provided any improvement in force control fidelity when applied during hard contact. The frequency

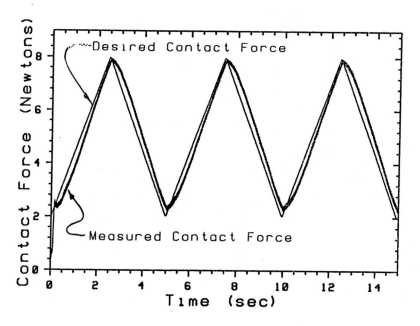

Figure A.7   Desired and Actual Force During Active Force Control

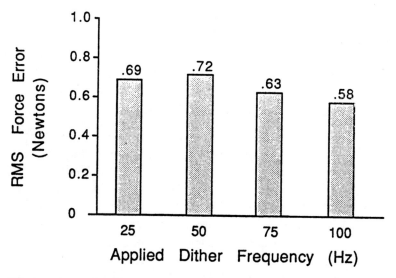

Figure A.8   RMS Force Error During Force Control with 2 Newton-meter
Dither Applied at each Frequency.

of the first bending mode goes from 20 rad/sec to 50 rad/sec when going
from the spring contact of the trials here to hard contact. It is apparent

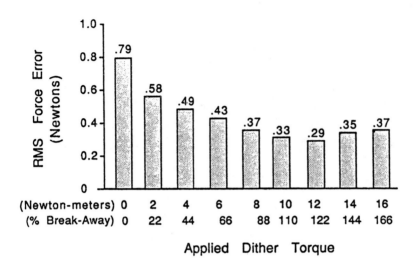

Figure A.9 RMS Force Error During Force Control with 100 Hz Dither
Applied at each Amplitude.

from the success of applying very large dither that the low pass filtering
effect of the mechanism and environment compliance is important. Which
leaves unsolved the challenge of the introduction: to control hard contact in
mechanisms with static friction.

A.4 Friction as a Function of Load

With the gliding experiment described in chapter 6.1, friction was
measured in joint one under three different load conditions. The results are
presented in table A.1. The applied torque was fourteen Newton-meters,
table lookup compensation was used and the mean gliding velocity was
measured five times under each load condition. The load torque presented is
the torsional load on joint 1 due to gravity, it acts around an axis orthogonal
to the direction of rotation, and thus does not effect the rotation torque
directly: the load is borne by the ball bearings. The equivalent additional
friction is a computed quantity equal to the amount of torque required to
cause the measured change in velocity given the joint viscous friction of 4.94
N-m per rad/sec.

Orthogonal loading gives a small but perceptible effect. The load
of 4 kg at full arm extension is nearly twice the manufacturer's specified

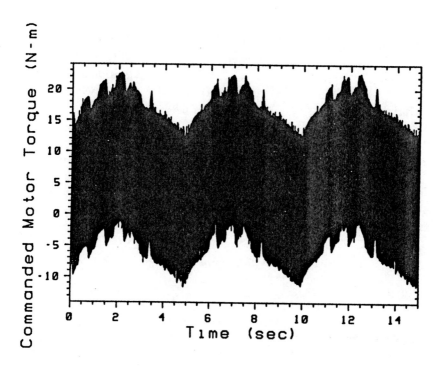

Figure A.10 Applied Motor Current during Force Control with 12 N-m of Dither at 100 Hz.

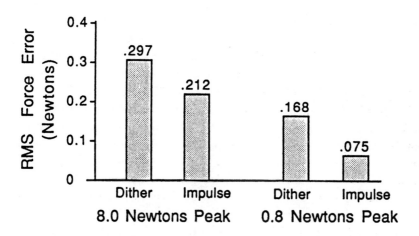

Figure A.11 Comparison of the RMS Force Error during Force Control Trials with Dither and the Impulsive Controller.

Table A.1   Measurements of Glide Velocity at Three Arm Loads.

Load Condition	Load Torque (N-m)	Glide Velocity (rad/sec)	Equivalent Additional Friction (N-m)
Arm Upright	2.3	1.029	-
Arm Extended	61	1.014	0.003
Arm Extended, 4 Kg Load	92	0.989	0.200

maximum carrying capacity of five pounds and results in an increase of 1.4% in the measured friction. We may conclude that bearing load is a minor contribution to friction.

A.5 Creep

Creep is the slow deformation of a material under a load insufficient to cause rapid failure. Ordinary glass, being a fluid, will creep under its own weight - at a rate of milli-meters per century. In control, the phenomenon of creep is of theoretical interest because it determines whether or not the friction - modeled as a function only of velocity, see section 2.4 - is discontinuous at zero velocity. A discontinuous friction force may violate the Lipschitz condition, which is a prerequisite for several important results, such as the small gain theorem.

The arm was allowed to come to rest and torque was applied to joint 1 in gradually increasing levels up to 8.0 N-m. The break-away data indicated that at 0.0022 radians, the position of the arm, break-away should occur at 9.1 N-m. As the torque was applied windup was observed. Over several trials, the mean windup was 0.0013 radian during the transition from 0 to 8.0 N-m of torque. From 8.0 N-m, the torque was increased in 0.1 N-m steps, one step per hour. The motor shaft encoder was used to detect motion; its resolution is 9964 counts per radian of arm motion. The results are presented in table A.2.

One shaft encoder count was recorded prior to full break-away: that observed during the hour at 8.5 N-m of torque. Windup was expected, but the occurrence of the count after 30 minutes of torque application suggests creep as the motion process. The absence of motion during the hours at 8.6 and 8.7 N-m shows the creep rate to be less than $1.1 * 10^{-8}$ radians per second with the applied torque greater than 97% of the break-away torque. The absence of motion for 18 hours at 8.0 N-m shows the creep rate to be

Table A.2   Raw Data Collected during 29 hours of Creep Investigation.

Torque (N-m)	Motion (radians)	Remarks
8.0	0	Torque Applied for 18 hours, 12 minutes.
8.1	0	
8.2	0	
8.3	0	
8.4	0	
8.5	0.0001	Motion occurred 30 minutes into the 1 hour trial.
8.6	0	
8.7	0	
8.8	1.875	Motion at moderate velocity to end of range, i.e., full break-away. Motion began 1 min. into trial.

less than $1.5 * 10^{-9}$ radians per second with an applied torque equal to 90% of the break-away torque. These results indicate that creep is exceedingly slow. A stability theory that requires a practical bound on the derivative of friction with respect to velocity is not supported by these data.

The effect of rapping was briefly investigated. With an applied torque of 8.2 N-m and the same initial position as above, the base of the robot was rapped lightly with a 1 oz. brass mallet. The strokes were applied in a way that would induce vibration in the drive train without directly applying torque. The rapping induced a steady motion of 1 shaft encoder count (0.0001 rad.) for each 10 raps. After 60 raps full break-away occurred and .2447 radians of motion. Note that the applied 8.2 N-m is substantially less torque than the expected 9.1 N-m break-away torque. The new stuck position corresponded to a location of high friction. Steady rapping apparently induces creep like behavior. Local vibration sources, such as equipment fans, may have contributed to the motion observed at 8.5 N-m of applied torque.

A.6 Effects that were not Observed

That which is not observed may be as important as that which is. In this section a few phenomena are discussed that were expected but did not make themselves apparent. The first of these is torque dependent friction. Because the normal force across some rubbing interfaces, notably the individual gear teeth, is affected by the applied torque, it was expected that a friction component proportional to motor torque would be in evidence. Such a friction component was long sought but never observed, though torque dependent friction is reported by DuPont, [90], and has been identified in the Stanford/JPL three finger hand [Fearing 87].

Magnetic Cogging is also observed in the Stanford/JPL hand, and was sought in the PUMA 560 experiments. But it was not observed. The experiment most sensitive to magnetic cogging is the measurement of static friction as a function of motor angle, shown in figure A.4. The spatial Fourier transform of this data, shown in figure A.12, shows a prominent feature at 31.25 cycles per motor revolution. The motor manufacturer (Magnetic Technologies) specifies a magnetic ripple at 25 cycles per revolution with a magnitude not greater than 4% of the applied torque. The feature at 31.25 cycles per revolution has a peak magnitude of 0.26 N-m per $\sqrt{\text{Hz}}$, and a square root power between the half maxima equivalent to a torque signal of 0.34 N-m. The magnitude of 0.34 N-m is 4.4% of the average torque of 8.0 N-m applied during the break-away experiment, putting the measurement in the range of the 4% upper bound set by the manufacturer. But the shift in frequency from the expected 25 cycles per revolution to 31.25 is unaccounted for.

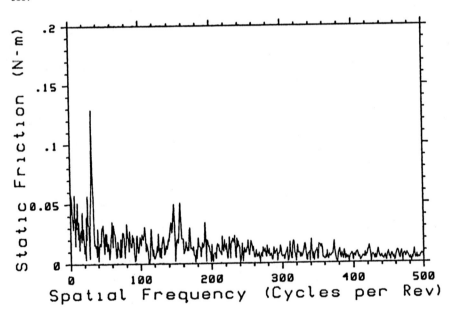

Figure A.12   Spatial Fourier Transform of the Static Friction of the Joint 2 Motor while Disconnected from Link 2.

Index

Printed in the United Kingdom
by Lightning Source UK Ltd.
9445900002B